Possibilities of Organic Farming in India

THE EDITORS

Dr. R. Chandra, Associate Professor, Department of Horticulture, A. S. College, Lakhaoti Bulandshahr (CCS University, Meerut, Uttar Pradesh). He did B. Sc. in Agriculture from R. B. S. College, Bitchpuri, Agra in 1999 and awarded degree of M. Sc. and Ph. D. in Horticulture with specialization in Floriculture and Landscaping in the year of 2002 and 2005 from GB Pant University of Agriculture & Technology, Pantnagar, U. S. Nagar (U. K.). He has qualified ICAR NET in Horticulture-Floriculture in 2004 and also cleared the exam of ARS in 2006. He has published 23 research papers in different research journals of national and international repute, more than 25 popular articles in different popular magazines, 01 book and 02 book chapters. He has participated and presented oral/poster presentations in more than 30 national and International seminars/symposiums/conferences. He is also a life member in national and international societies. He received the Excellence Teaching Award in 2020 from Agriculture Technology Development Society (ATDS), Ghaziabad.

Dr. Sachin Kumar Jain was born on December 21st, 1990 in Shamli district, Uttar Pradesh, India. He did his B.Sc. (Ag) from R.K. (PG) College, Shamli affiliated to C.C.S. University, Meerut (U.P.) in 2011 and awarded with Vice Chancellor Gold Medal for received Ist merit rank at University level. He did his M.Sc.(Ag.) in Plant Pathology (2013) and Ph. D. Plant Pathology (2016) from Department of Plant Pathology, Sardar Vallabhbhai Patel University of Agriculture & Technology, Modipuram, Meerut (U.P.). He joined in Plant Pathology department of Amar Singh (PG) College, Lakhaoti as an Assistant Professor (Plant Pathology) in 2017. Dr. Jain has published more than 22 research papers in national journals. He has published more than 15 abstracts in national and international conference souvenir and 20 popular articles in various magazines. He is a life member of National Science Congress and Indian Phytopathological Society.

Dr. Raj Pal Singh obtained his B.Sc.(Ag) Degree from Narendra dev University of Agriculture and technology, Kumarganj , Faizabad, M.Sc(Ag) in Agronomy from Banaras Hindu University, Varanasi, Uttar Pradesh ,Ph.D in Agronomy from C.C.S.University , Meerut and NET from ICAR . He had professional experience in CAPART Ministry of Rural Development as young Professional (2005-2006) , as SMS Agronomy in GVT-KVK, Godda, Jharkhand from October 2006 to November 2017 in 24 November 2017 Appointed as Assistant Professor in A.S. college, Lakhaoti, Bulandshahr. He had Total 13 research papers, 25 popular articles, 03 Training manuals, 02 books.He got ICAR award, Vice chancellor award and several other awards by different organizations.

Dr. Ajay Kumar is Assistant Professor in the department of Plant Pathology, Amar Singh (P.G.) College, Lakhaoti, Bulandshahr, Uttar Pradesh under the CCS University Meerut. He did his B.Sc. (Agriculture) in 2009 from Udai Pratap Autonomous College, Varanasi, U.P.; M.Sc. (Plant Pathology) in 2011and Ph.D. (Plant Pathology) during 2015 from Acharya Narendra Deva University of Agriculture and Technology, Ayodhya, U.P. He has cleared ICAR-NET and also awarded with Rajiv Gandhi National Fellowship by UGC during Ph.D. programme. He has published **30** research papers&**1** review paper in national and international journals, **9** book chapters, **5** books, **2** newsletters and **15** popular articles. He supervised **1** student of M.Sc. Plant Biotechnology. Dr. Kumar has joined the life membership of **3** societies and received **4** awards by different societies, served the duties as joint-organizing secretary in the International Conference and delivered the guest lectures on mushroom cultivation.

Possibilities of Organic Farming in India

– Editors –

R. Chandra

S. K. Jain
(Corresponding Editor)

R. P. Singh

Ajay Kumar

2023

Daya Publishing House®
A Division of
Astral International Pvt. Ltd.
New Delhi – 110002

ISBN 9789359191645

Published by : **Daya Publishing House®**
A Division of
Astral International Pvt. Ltd.
– ISO 9001:2015 Certified Company –
4736/23, Ansari Road, Darya Ganj
New Delhi-110 002
Ph. 011-43549197, 8130496929
E-mail: info@astralint.com
Website: www.astralint.com

Preface

Organic farming is an agricultural process that uses pest control derived from organic manure and animal or plant waste. Synthetic chemical and pesticides are not used on organic and natural farming. Apart from enhancing soil quality and fertility, the natural farming practices are also beneficial for the rural economy as cattle are an important part of such farming practices. Organic farming is in a nascent stage in India. According to the Union Ministry of Agriculture and Farmers' Welfare, about 2.78 million hectare of farmland was under organic cultivation as of March 2020. This is two per cent of the 140.1 million ha net sown area in the country.

A few states have taken the lead in improving organic farming coverage, as a major part of this area is concentrated only in a handful of states. Madhya Pradesh tops the list with 0.76 million ha of area under organic cultivation — that is over 27 per cent of India's total organic cultivation area. The top three states — Madhya Pradesh, Rajasthan and Maharashtra — account for about half the area under organic cultivation. The top 10 states account for about 80 per cent of the total area under organic cultivation. Modern agriculture and food systems, including organic agriculture, are undergoing a technological and structural modernization and are faced with a growing globalization. Organic farming can be seen as pionering efforts to create sustainable development based on other principles than mainstream agriculture. There are however large differences between the challenges connected to, on the one hand, modern farming and consumption in high-income countries and, on the other hand, smallholder farmers and resource-poor consumers in low-income countries. The point of departure is the increasing globalization and the production and trade of food and fodder and how this influences the role of organic farming. The main aim of this book is to provide an overview of the potential role and challenges of organic farming in this global perspective, as seen from different perspectives such as sustainability, food security and fair trade. Initially, the book provides an overall status of global trends in agriculture followed by discussions of possibilities sustainability, globalization and the relative new concepts of 'ecological justice' and 'political ecology'. Different views on economy and trade are furthermore discussed with a focus on ecological economics. Then, the status and possibilities of organic farming in India is discussed, including problems of nutrient cycles and soil depletion plus issues on veterinary medicine. Furthermore, organic farming is related to the world food supply. The possibilities of knowledge exchange in organic farming are also evaluated and it is assessed how a large scale conversion to organic farming would impact on food security. Finally, prospects and challenges of organic farming in India are discussed in a synthesis chapter. Readers who seek first an overview and summary across the different chapters are recommended to start by reading the synthesis.

Finally, we would want to give special thanks to the publishers and printers who have worked tirelessly to ensure that this book is in perfect shape and available to students on time.

R. Chandra

S.K. Jain
R.P. Singh

Ajay Kumar

Contents

Chapter 1

Indian Organic Farming: Its Concept, Components and Challenges

Ajay Kumar[1] *and Subhash Chandra*[2]

[1]*Department of Plant Pathology, Amar Singh College (CCS University Meerut), Lakhaoti, Bulandshahar, Uttar Pradesh*
[2]*Department of Plant Pathology, Acharya Narendra Deva University of Agriculture and Technology, Kumarganj, Ayodhya, Uttar Pradesh*
e-mail: kaushal4980@gmail.com

ABSTRACT

Organic farming includes the concept that the soil, plant, animals, and humans are linked. Organic farming encourages the use of crop rotations and use of organic manure including green manure for increasing soil organic matter. Carefully management of soil offers some advantages with respect to the increase water holding capacity, ion exchange and reduces soil erosion. Green manuring and inter-cropping of legumes is another important aspect for organic farming systems. It is not only helps in controlling weeds but also in improving its chemical and physical properties by reducing the leaching of nutrients and reducing soil erosion. Crop rotation and green manuring are known to be improving soil fertility.

Keywords: *Green manuring, Fertility, Challenges, History, Components.*

INTRODUCTION

Organic farming is defined as a production system that avoids or largely eliminates the usage of synthetic compounded fertilizers, growth regulators, pesticides, and farm animal feed additives. To maximize feasible extent, the organic farming system depends on crop rotation, green manures, legumes, animal manures, crop residues, off-farming natural squanders, and aspect of biological pest control to preserve soil fertility and productivity to sustain plant and thereby curbing pests, diseases, and unwanted weeds. Organic farming methods are internationally regulated and legally executed by many countries, based in great part on the standards set by the International Federation of Organic Agriculture Movements (IFOAM) an international umbrella organization for organics established in 1972.

India is an Agrarian country with around 60 per cent of its people directly or indirectly depends upon agriculture. In archaic time the practice of agriculture was considered to be a maximum service to the society and this practice was inter-twined in their convention and culture.

History of Organic Farming

- The conviction of organic farming developed in the early 1900s by Sir Albert Howard, F.H. King, Rudolf Steiner, and others who believed that the use of animal manures (often made into compost), cover crops, crop rotation and biologically based pest controls resulted in a better farming system.
- India as an agricultural researcher, gained much inspiration from the traditional and sustainable farming practices he encountered there and advocated for their adoption in the West. Such practices were further promoted by various advocates-such as J.I. Rodale and his son Robert, in the 1940s and onward, who published *Organic Gardening and Farming* magazine and a number of texts on organic farming. The demand for organic food was stimulated in the 1960s by the publication of *Silent Spring*, by Rachel Carson, which documented the extent of environmental damage caused by insecticides.
- Organic agriculture is not alien to India. In fact, the first scientific approach to organic farming dates back to the Vedas of the later Vedic period, the essence of which is to live in harmony with, rather than exploit, Mother Nature. There is brief mention of several organic inputs in our ancient literatures like Rigveda, Ramayana, Mahabharata, Kautilya Arthasashthra *etc.* In fact, organic agriculture has its roots in traditional agricultural practices.
- Ancient period Oldest practice 10000 years old, dating back to Neolithic age, practiced by ancient civilization like Mesopotamia, Hwang-Ho basin *etc.*
- Ramayana All dead things - rotting corpse or stinking garbage returned to earth are transformed into wholesome things that nourish life. Such is the alchemy of mother earth as interpreted by C. Rajagopalachari

- ✰ Mahabharata (5500 BC) Mention of Kamadhenu, the celestial cow and its role on human life and soil fertility.
- ✰ Kautilya Arthashastra (300 BC) Mentioned several manures like oil cake, excreta of animals.
- ✰ Brihad-Sanhita (by Varahmihir) described how to choose manures for different crops and the methods of manuring.
- ✰ Rig Veda (2500–1500 BC) Mention of organic manure in Ria Veda 1, 161, 10, 2500-1500 BC, is Green Manure in Atharva Veda II 8.3, (1000 BC). In Sukra (IV, V, 94, 107-112) it is stated that to cause healthy growth, the plant should be nourished by dungs of goat, sheep, cow, water as well as meat. A reference of manure is also made in Vrksayurveda by Surpala.

Concept of Organic Farming

Organic farming includes the concept that the soil, plant, animals, and humans are linked. Concept of organic farming is based on some below mentioned principles:

1. Nature is considered as a best role model for farming, since it does not use any inputs nor demand unreasonable quantities of water.
2. Organic farming system does not believe in mining of the soil of its nutrients and protect the long term fertility of soil.
3. The soil in this system is considered as living entity.
4. On sustained basis living population of microbes and other organisms in soils are significant contributors to its fertility.
5. Development of biological diversity and the maintenance and replenishment of soil productivity is the major concern of organic farming systems.
6. Importance given on crop rotation, natural predators, biological pest, disease and weed management, resistance varieties instead of chemical control.

Components of Organic Farming

1. Crop and Soil Management

Organic farming encourages the use of crop rotations and use of organic manure including green manure for increasing soil organic matter. Carefully management of soil offers some advantages with respect to the increase water holding capacity, ion exchange and reduces soil erosion. Green manuring and inter-cropping of legumes is another important aspect for organic farming systems. It is not only helps in controlling weeds but also in improving its chemical and physical properties by reducing the leaching of nutrients and reducing soil erosion.

2. Organic Manures

Organic material such as FYM, biogas slurry, compost, straw, crop residues, green manures and cover crops are good substitutes for inorganic fertilizers. Similarly, sea weeds, fish manure, basic slag and rock phosphate also proved better. Organic material helps in improving physio-chemical properties of soil. It reduces soil erosion. Crop rotation and green manuring are known to be improving soil fertility.

3. Non-chemical Weed Management

Weed management is one of the main concerns in organic agriculture. The some important practices that consider in preventing weed problems are tillage, crop rotation, and manure management, use of cover crops, mulches and green manuring.

4. Domestic and Industrial Waste Recycling

Among the industrial by products, molasses and press mud from sugar industry have good manurial value. It is important to use only well decomposed press mud at 10 tones/ha. Addition of press mud improves the soil fertility and enhances the activity of microbes. Coir waste can be used as manure after proper decomposition. Use of sewage and sludge for crop production can form an important component of organic farming.

5. Biological Pest Management

Control of insect pests and pathogens is a challenging job in organic farming. The conservation of natural enemies of pests is important for minimizing the use of chemical pesticides and for avoiding multiplication of insecticide resistant pests. Botanical pesticides derived from neem as well as parasitoids like *Trichoderma, Bacillus thuringiensis* and NPV (Nuclear Polyhedrosis Virus) are used in pest management.

Challenges of Indian Organic Farming

There are two micro and macro level challenges that must be overcome to achieve Indian Organic Farming.

1. Micro Level Issues

The micro level issues confronting organic farming include economic viability, particularly for small and marginal farmers, marketing, *etc.* For example, one of the greatest barriers for organic farming is the so-called conversion period due to the direct and indirect costs. The conversion of a conventional farm to an organic farm requires strictly adhering the rules and standards of production, processing and labelling at prescribed international levels. During the conversion period all the standards required for certifying a product as 'organic' must be fulfilled and verified by a certifying agency. Costs due to things, such as information, marketing charges, inspection, and certification expenses also increase the cost of organic farming. For instance, fees for the inspection and certification can be prohibitively high at Rs. 5000, since this equals the returns from agriculture for many small farmers (Brook and Bhagat, 2004).

2. Macro Level Issues

The macro challenges include impacts mainly on food security, employment, and environment. The question of food security assumes significance considering potential yield reductions of organic farming vis-à-vis conventional farming (Pandey and Singh, 2012), particularly in the two to three year conversion period. Considering the various challenges to the adoption of organic farming the Working Group on Organic and Bio-dynamic Farming of the Planning Commission suggested examination of some important issues for effective promotion and practice of organic farming and sustainable agriculture. These include economics of organic crop production, economic and environmental externalities associated with conversion to organic farming, comparative study of chemical based and organic farming covering social, environmental and economic costs.

Organic Farming for Sustainable Agriculture

Sustainable is derived from the Latin word, *sustinere,* meaning to keep in existence, implying permanence or long-term support. In the context of agricultural production, according to Ikerd (1993), sustainable agriculture as 'capable of maintaining its productivity and usefulness to society over the long run, it must be environmentally-sound, resource-conserving, economically viable and socially supportive, commercially competitive, and environmentally sound". Sustainable agriculture integrates three main goals profitability, and social and economic equity.

1. Organic Source of Plants Nutrients

At present, most optimistic estimates show that about 25-30 per cent of nutrient needs of Indian agriculture can be met by various organic sources. Supplementation of entire nitrogen through Farm Yard Manure (FYM) sustains crop productivity at more than use of conventional nitrogen fertilizers. Since the apprize of NPK availability from organic sources are based on total nutrient content, efficiency of these sources to meet the nutrient requirement of crops is not as assured as mineral fertilizers, but the joint use of chemical fertilizers along with various organic sources is capable of sustaining higher crop productivity, improving soil quality, and productivity on long-term basis.

2. Organic Nutrition on Crop Productivity

The increase of organic matter in the soil is a well-known practice to increase crop yields. Sharma and Mitra (1990) reported that the application of organic materials increased grain and straw yield of rice. Ranganathan and Selvaseelan found that application of spent mushroom and rice straw compost though comparable with FYM increased rice grain yields by 20 per cent over NPK fertilizer. Singh *et al.* reported that the application of 7.5t FYM ha^{-1} produced significantly more grain, and straw yields over unfertilized fields. The entire yield attributing characters of rice increased with increasing rates of FYM. Organic farming with dhaincha (*Sesbania aculeata* L.) made considerable improvement in grain yield of rice and Chickpea.

3. Organic Nutrition on Soil Fertility

Minhas and Sood (1994) also reported that the organic matter after decomposition release macro- and micronutrients to the soil solution, which becomes available to the plants, resulting in higher uptake. Organic farming was capable of sustaining higher crop productivity and improving soil quality and productivity by manipulating the soil properties on long term basis. It was reported that organic and low-input farming practices after 4 years led to an increase in the organic carbon, soluble phosphorus, exchangeable potassium, and pH and also the reserve pool of stored nutrients.

4. Organic Nutrition on Soil Biological Properties

Compost contains bacterial, actinomycetes, and fungi; hence, a fresh supply of humic material not only added microorganisms but also stimulated them. Besides, compost played an important role in control of plant nematodes and in mitigating the effect of pesticides through sorption. Sorption is the most important interaction between soil/organic matter and pesticides and limits degradation as well as transport in soil. Pesticides bound to soil organic matter or clay particles are less mobile, bio-available but also less accessible to microbial degradation and thus more persistent.

Aims and Objectives of Organic Farming

- ☆ Provide food of high nutritional quality in sufficient quantity.
- ☆ To work with natural systems rather than dominating them.
- ☆ To encourage and enhance the biological cycles within farming system involving micro-organisms, soil flora and fauna, plants and animals.
- ☆ To maintain and increase the long term fertility of soils.
- ☆ Avoid all forms of pollution that may result form agricultural techniques.
- ☆ Maintain the genetic diversity of the agriculture system.

- To consider the wider and ecological impact of the farming system.
- To maintain the ecological balance.

Characteristics of Organic Farming

- Maximum but sustainable use of local resources.
- Minimum use of purchased inputs.
- Ensuring the basic biological functions of soil, water, nutrients and humus.
- Maintaining a diversity of plant and animal species as a basis for ecological balance and economic stability.
- Increasing crop and animal diversity in the form of agro-forestry systems, integrated crop/livestock systems *etc.*

Advantages of Organic Farming

- Organic manures produce optimal conditions in the soil for high yield and good crop quality.
- They supply primary, secondary and micro-nutrients to crops.
- They improve growth and physiological activities of plants.
- They improve physio-chemical properties of soil.
- Addition of organic carbon in the organic matter is the source of energy for microbes which help in better soil aggregation.
- They reduce need of purchased inputs.
- Minimizes pollution due to waste material.
- Organic fertilizers are considered as complete plant food.
- Organic matter restores the soil pH.
- Organically grown crops produced nutrition and healthy food free from toxic residues.
- Organically grown plants are more resistant to disease and insect pest.
- There is an increase in consumers demand for organic produce.
- Organic farming prevents soil as well as environmental degradation.
- Organic farming helps to regenerate degraded lands.
- Due to diversification of crops there is more secured income.
- It minimizes the cost of production through use of on-farm inputs.
- Quality of food material is safe without any health hazards.

Disadvantages of Organic Farming

- In changing over to organic farming initial crop loss generally occurs.

- There are no fully developed markets for organic products.
- Biological controls may have weakened or destroyed due to organic chemicals.

CONCLUSION

Organic material such as FYM, biogas slurry, compost, straw, crop residues, green manures and cover crops are good substitutes for inorganic fertilizers. Similarly, sea weeds, fish manure, basic slag and rock phosphate also proved better. Organic material helps in improving physio-chemical properties of soil. Carefully management of soil offers some advantages with respect to the increase water holding capacity, ion exchange and reduces soil erosion. It reduces soil erosion. Green manuring and inter-cropping of legumes is another important aspect for organic farming systems. It is not only helps in controlling weeds but also in improving its chemical and physical properties by reducing the leaching of nutrients and reducing soil erosion. Crop rotation and green manuring are known to be improving soil fertility. Organic farming encourages the use of crop rotations and use of organic manure including green manure for increasing soil organic matter.

REFERENCES

Archna, K. (2013). Role of Indian Government in the Development of Organic Agriculture, *IOSR Journal of Agriculture and Veterinary Science* (IOSR-JAVS), e-ISSN: 2319-2380, p-ISSN: 2319-2372.2 (6): 32-39.

Balasubramanian, A., Siddaramappa, R. and Rangaswami, G. (1972). "Effect of organic manuring on the activities of the enzymes hydrolysing sucrose and urea and on soil aggregation, " Plant and Soil, 37(2): 319-328.

Bhattacharyva, P. and Chakraborty, G. (2005). Current status of organic farming in India and other Countries, *Indian Journal of Fertilisers*, 1(9): 111-123.

Brook, and Bhagat, G. (2004).Organic by default: The irony of organic farming in India, available at: http: //www.ecoworld.org/Home/Articles2.cfm?TID=349.

Chhonkar, P.K. (2002). "Organic farming myth and reality, " in Proceedings of the FAI Seminar on Fertilizer and Agriculture Meeting the Challenges, New Delhi, India.

Clark, M.S., Horwath, W., Shennan, R.C. and Scow, K.M. (1998). "Changes in soil chemical properties resulting from organic and low-input farming practices, " *Agronomy Journal*, 90(5): 662-671.

Doran, J.W. and Zeiss, M. (2000). "Soil Health and Sustainability: Managing the Biotic Component of Soil Quality." *Applied Soil Ecology* 15 (1): 3-11.

Ettema, Christien, H. (1998). "Soil Nematode Diversity: Species Coexistence and Ecosystem Function." *Journal of Nematology*, 30(2): 159-69.

Gaur A.C. and Prasad, S.K. (1970). "Effect of organic matter and inorganic fertilizers on plant parasitic nematodes," *Indian Journal of Entomology*, (32): 186-188.

Gaur, A.C., Sadasivam, K.V., Vimal, O.P. Mathur, R.S. and Kavimandan, S.K. (1973). "Studies on the humification of organic matter in a red Rakar soil, " Zentralblatt fur Bakteriologie, 128(1): 149-161.

Ikerd J. (1993).Two related but distinctly different concepts: organic farming and sustainable agriculture. Small Farm Today, 10(1): 30-31.

Kulhade, A., Gupta, A.D., Mishra, A. and Singh, S.R.K. (2016).Role of organic farming in Indian agriculture. *Flora and fauna*, 22(1): 41-48.

Meena, R.K., Meena, R.S., Naik, B.S.S.S., Meena, B.L. and Meena, S.C. (2020). Organic farming-concept, principles, goals and as a sustainable agriculture: A review, *International Journal of Chemical Studies*, 8(4): 24-32.

Meena, R.K., Neupane, M.P. and Singh, S.P. (2015).Effect of phosphorus levels and bio organic sources on growth and yield of rice (*Oryza sativa* L.). *International journal of Agricultural Science*, 11(2): 26-289

Minhas, R.S. and Sood, A. (1994)."Effect of inorganic and organic on yield and nutrients uptake by three crops in rotation in aid alfisol, " *Journal of the Indian Society of Soil Science*, 42: 27-26.

Pandey, J., Singh, A. (2012).Opportunities and constraints in organic farming. An Indian perspective. *Journal of Scientific Research*, 56: 47-72.

Ranganathan, D.S. and Selvaseelan, D.A. (1997). "Mushroom spent rice straw compost and composted coir pith as organic manures for rice, " *Journal of the Indian Society of Soil Science*, 45(3): 510-514.

Rigby, D. and Aceres, D. (2001). Organic farming and the sustainability of agricultural systems. *Agricultural Systems*, 68: 21-40.

Sharma, A.R. and Mitra, B.N. (1990). "Complementary effect of organic material in rice-wheat crop sequence, " *The Indian Journal of Agricultural Sciences*, 60(3): 163-168.

Singh, G., Singh, O.P., Yadava, R.A. Singh, P.P. and Nayak, R. (1998)."Response of rice (*Oryza sativa*) varieties to nitrogen levels in flash flood conditions, " *Indian Journal of Agronomy*, (43)3: 506-510.

Singh, G.B. and Dwivedi, B.S. (1996). Integrated nutrient management for sustainability. *Indian farming*, 46(9): 9-15.

Yadav, S.K., Babu, Subhash, Yadav, M.K., Singh, Kalyan, and Yadav, G.S. (2013). A Review of Organic Farming for Sustainable Agriculture in Northern India. *International Journal of Agronomy*, (718145): 8.

Chapter 2

Organic Farming in India

Ajendra Kumar[1], D. Ram[2] and Ranjeet Kumar[1]

[1]Research Scholar; [2]Professor,
Acharya Narendra Deva University of Agriculture and Technology Kumarganj, Ayodhya–224229, Uttar Pradesh

"With wrong farming methods, we turn fertile land into desert.

Unless we go back to organic farming and save the soil, there is no future"

— Jaggi Vashudev

ABSTRACT

Indian economy is basically based on agriculture about 60 per cent population of India are engaged in agronomic practices who fill the need of food for the whole country as the population increasing steadily. The size of land holding goes down triggers the farmers to use more and more fertilizers, weedicides and pesticides to improve yield but excess of the chemicals not only pollute the atmosphere but also a major cause of several death causing disease like cancer etc. Due to these problems food quality and safety are the vital factors which has attained constant attention in common people. Increasing environmental awareness and frequent food hazards highly decreased the trust towards food quality. In recent years organic practices have gained increasing popularity and ecologically grown food has become one of best choices for both consumers and growers and is a part of green life style. In recent past years, the organic farming practices across the India is increasing but the magnitude through which the farming system should be reform is very low. National Program of Organic Production (NPOP), Rastriya Krishi

Viklas Yojna (RKVY), Pramparagat Krishi Vikas Yojna (PKVY) plays very important role in dissemination of awareness and providing financial assistance to develop organic techniques which are beneficial for both environment and human heath

Keywords: *Organic farming, Weedicide, Pesticide, Ecology, RKVY, PKVY, NPOP.*

INTRODUCTION

Organic farming is an agricultural system that uses fertilisers of organic origin such as compost manure, green manure and bone meal and place emphasis on techniques such as crop rotation and companion planting. India ranks 8th in terms of World's Organic Agricultral Land and 1st in terms of numbers of producers as per data (Source: FIBL and IFOAM Years Book, 2020). APEDA (Agricultural and Processed Product Export Development Authority), Ministry of Commerce and Industries is implementing the National Programme for Organic Production (NPOP). The programme involves the accreditation of certification bodies, standards for organic production promotion of organic farming and marketing *etc.* The NPOP standards for the production and accreditation system have been recognised by European commission and Switzerland for unprocessed plant products as equivalent to country standards. With these recognitions Indian organic products duly certified by the accredited and certification bodies of India are accepted by importing countries.

Status of Organic Farming in India

In recent days awareness among the farmers regarding the organic farming is continuously increasing. In India, about 2.78 million hectare of farmer land was under organic cultivation (Union Ministry of Agriculture and Farmer's welfare report, March 2020). Madhya Pradesh has maximum area (0.76 million hec.) under the organic farming which contributes over 27 per cent of total organic cultivation area. Sikkim is the only state which is fully organic. Majority of states have less contribution in organic farming. Overall area under organic system is only 2 per cent of net sown area in India which is very less as compared to other countries. Currently only 12 states *i.e* Madhya Pradesh, Gujrat, Telangana, Sikkim, Bihar, Karnatka, Uttrakhand, Chattisgarh, Tamil Nadu and Uttar Pradesh have their own certification agency accredited by Agricultural and Processed Food Products Export Development Authority (APEDA).

India introduced the organic farming policy in 2005. Several government schemes in India intersperse the awareness among the farmers for benefits of Organic cultivation *viz.* National Programme for Organic Production (NPOP) was launched in 2000 enhance the area of organic cultivation by 1.94 million ha which is nearly 70 per cent of total area cultivated under organic farming in India. Major Initiatives which played significant role in promotion of organic farming in India are listed as:

Programmes	*Highlights*
National Project on Organic Production (NPOP)	Central sponsored scheme launched in 2000 with an outlay of rupees 57.04 crore which was further expanded in 11th five year plan with outlay of 101 crore rupees **Objective:** ☆ To encourage the organic food production, promote manufacture and usage of organic and biological inputs, such as organic manure, bio-control agents and bio-fertilizes ☆ Provides financial aid upto 63 lakhs (33 per cent of capital cost) for constructing fruits and vegetables waste compost units, through NABARD ☆ Provides subsidy upto 40 lakhs (25 per cent of the capital cost) for the construction of bio-fertilizer or bio-pesticide production unit through NABARD
National Project on Management of Soil Health and Ferility (NPMSF)	Implemented during the 11th FYP with an oultlay of 429.85 crore rupees **Objective:** ☆ To promote the judicious and balanced use of fertilizers and organic manure on the basis of soil test results ☆ Provides financial assistance of rupees 500/ha for promoting the use of organic manure
Network Project on Organic Farming (NPOF)	Initiated by ICAR in 2004-2005 at the Project Directorate for Farming system Research Farm at Modipuram **Objective:** ☆ To develop packages and practices for different crops and farming systems for organic production in different agroclimatic conditions of India ☆ Packages and practices have been developed for basmati rice, rainfed wheat, maize, red gram, chick pea, soyabean, ground nut, mustard, isabgol, black pepper, ginger, tomato, cabbage and cauliflower

Programmes	*Highlights*
National Horticulture Mission (NHM) and Horticulture Mission for North East and Himalayan State	Centrally sponsored scheme, launched in 2005-06 **Aim:** ✰ To strengthen the growth of the horticulture sector comprising fruits, vegetable, roots and tuber crops, mushroom, spices, flowers, aromatic plants, cashew and cocoa. ✰ Provides subsidy of 50 per cent for establishing vermin-compost units and HDPE vermin-beds ✰ Provide assistance for the organic certification of 5 lakh for a group farmers covering an area of 50 hectares ✰ Provides ₹30000 per beneficiary for adopting organic farming.
Rastriya Krishi Vikas Yojana (RKVY)	**Objective:** ✰ Provides assistance to the projects formulated and approved by the state for the decentralized production and marketing of organic fertilizers
National Mission for Sustainable Agriculture (NMSA)	**Objective:** ✰ 100 per cent assistance the state governments for setting up mechanization of fruit and vegetable waste. ✰ 100 per cent aid for the setting up quality control for testing bio-fertilizers upto 85 lakh
Paramparagat Krishi Vikas Yojna (PKVY)	**Objective:** ✰ Provides ₹20,000 to farmers upto 3 years for performing organic farming ✰ Procuring packaging material, preparation of labels, holograms, printing and branding of organic produce at 2,500/acre ✰ Provides financial aid for a cluster of 50 acres, to the tune of ₹1,20,000 for transporting organic produce to the market place. ✰ Financial assistance of ₹36,330/cluster to motivate and support marketing facilities is provided to organize an organic fair

Organic Farming Process

Organic farming system basically depends upon the development of social, economic and ecological food production system. The main principle and practices of organic food to inspire and enhance the biological cycle in the farming system, keep and improve deep rooted soil fertility, reduces all types of pollution, evade the use of fertilizers and pesticide, conserve genetic diversity in food, consider the vast social ecological impact of food production in sufficient quantity. To get the certification of organic produce without the use of any synthetic chemicals for 3 years before harvesting with enough buffer zone to lower contamination from the adjacent farms. Genetically engineered products, sewage sludge, and ionising radiation are strictly prohibited in the organic production fertility and nutrient management through framing practices *viz.* crop rotation, organic mulches, biological control of disease and pest. Organic livestock should be reared devoid of scheduled application of growth hormones for antibiotics and they should provide enough access to outdoor. Preventive health practices such as routine vaccination, vitamins and minerals supplementation are also needed (Organic Food Production act, 1990).

Aims of Organic Farming in India

The four aims of organic farming are:

- ☆ **Principles of Health:** health of ecosystem, people and communities.
- ☆ **Principle of Ecology:** right balance between ecosystem and environment.
- ☆ **Principles of Fairness:** good human relationships and quality of life.
- ☆ **Principle of Care:** consideration about environment in present and future.

Techniques of Organic Farming

There are some techniques by which organic farming in India is practised that not only improve yield but also enhance the quality of produce

1. **Soil Management:** After the harvesting of produce soil lose its nutrients and its nutrient availability capacity goes down. The process in which soil is recharges with all essential nutrients is called soil management. Organic farming utilises natural resources to increase soil fertility. It uses bacteria available in the animal waste making soil more productive and fertile
2. **Weed Management:** The main of organic farming to remove the weeds. Weeds are unwanted plants growing with crop. Competition among the weeds and crop heavily reduces the crop yield. The techniques which resolve the problem of weeds are (i) moving or cutting in this

process cut the weed and (ii) Mulching; here farmers use a plastic film or plant residue on the soil surface to block the weeds growth.

3. **Crop Diversity:** In this technique different crops can be cultivated together to meet the growing demands for crops
4. **Biological Pest Control:** Application of living organism *viz.* predators to control the pest without use of chemicals.

Organic Farming Methods

There is a set of minimum requirements to certify any farm as organic. They are:

1. **Conversion Period:** If the farmers are practising conventional farming and field is not fully organic then the farmer must have a conversion plan. The interval between the beginning of organic operations and certification is the conversion period. The exact time period is decided taking into consideration the ecology of land and its previous use.
2. **Mixed Farming:** integration of animal use in the farming system to ensure the soil fertility and crop yield.
3. **Cropping Pattern:** Growing the same crop in the field repeatedly must be avoided as this would affect the soil fertility. Cultural practices as inter cropping would help soil maintain its vitality. The practices of inter cropping helps in disease and insect pest control effectively.
4. **Planting in Organic Farming:** The planting material and seeds must be procured from a reliable source that is certified organic by the authorizing body. In case organic ones are not available then chemically untreated plant material should be used. Pollen culture seeds, tissue culture seeds, transgenic plants, genetically engineered seeds are also not allowed
5. **Manuring Requirements:** The fertility of soil must be maintained through the application of leguminous crops, green manure, bio-degradable wastes are used as manures. Crop and animal residues must be recycled back into soil directly or indirectly. Manure like farm yard manure, vermi compost *etc.* are allowed but chemical fertilizers are not.
6. **Pest and Disease Management:** Natural enemies of pest are reared to protect the crop. Plant based repellents, neem seed kernel extracts, mechanical traps, pheromone traps, clay, soft soap and chromatic traps are permitted for use in farms

Regulatory Body and Certification

To improve the quality of organic produce regulatory body at state level and central level existed

- ✰ National Program on Organic Production (NPOP) is the authority defining the regulatory mechanism. Two different bodies for the regulation of in the domestic and export market: (i) Foreign Trade Development Regulation Act, NPOP looks after the export requirements; (ii) Agriculture Produce Grading, Marketing and Certification Act, it looks after the domestic market and import
- ✰ Agricultural and Processed Foods Export Development Authority (APEDA) is a regulatory body of NPOP as per FTDR Act. It is also the agriculture marketing Advisory under Agriculture Ministry.

CONCLUSION

Organic farming in India is a safe venture although it may take a while to establish and become fully functional. There are government subsidies available for this and if this method is strictly followed with the required organic certification, then organic farming in India is very profitable.

REFERENCE

Das, S., Chatterjee, A. and Pal, T. K. (2020) Organic Farming in India: a vision towards a healthy nation. *Food Quality and Safety Oxford Academic*, 4(2): 69-76

Front. Sustain. Food Syst. (2019) URL https: //doi.org/10.3389/fsufs. 2019.00082

Gill, M. S., Prasad, K. (2009). Network Project on Organic Farming-Research highlights. Organic Farming Newsletter, 5(2): 3-10.

Mahanta, D., Bhattacharya, R., Gopinath, K. A. and Tuti, M. (2013) Influence of farmyard manure application and mineral fertilization on yield sustainability, carbon sequestration potential and soil property of garden pea–french bean cropping system in the Indian Himalayas. *Scientia Horticulturae* 164: 414-427

Peramaiyan P., Hermanseen, J. E. and Halberg, N. (2011) Food Security of Small Holding Farmers: Comparing Organic and Conventional Systems in India. *Journal of Sustainable Agriculture.* 35(1): 48-68

Present status and prospects of organic farming in India, *European Academic Research,* vol.3, dated July 15, URL: https: //www.researchgate.net/publication/282243181_Present_Status_and_Prospects_of_Organic_Farming_in_India

Saha, S.; Mina, B.; Gopinath, K. A.; Kundu, S. (2008) Relative changes in phosphatase activities as influenced by source and application rate of organic composts in field crops. *Bioresource Technology,* 99(6): 1750-1757

The world of Organic Agriculture: Statistics and Emerging Trends, 2020, *FiBL and IFOAM, Organics International,* Pg no. 51, 58, 66, 191.

Chapter 3

Prospects and Constraints in Organic Farming

Shivanand Maurya[1], Ravi Verma[1] and Ashok Kumar[2]

[1]Department of Agronomy, Acharya Narendra Deva University of Agriculture and Technology, Kumarganj, Ayodhya, Uttar Pradesh
[2]Department of Plant Pathology, Post Graduate College, Ghazipur, Uttar Pradesh
e-mail: kaushal4980@gmail.com

ABSTRACT

India has traditionally practiced organic agriculture, but the advent of green revolution technologies led to the increased use of chemicals. Presently there is lack of government subsidies or support to make conversion to organic status easier and cheaper. The question about yield and financial viability of organic farming are crucial and there are no empirical studies available in the Indian context comparing the economic and ecological returns of organic farm vis-à-vis conventional farms. Organic agriculture has been neglected in the agriculture policy, and therefore there is less government assistance for the promotion of organic agriculture in the form of subsidies, agricultural extension services and official research.

Keywords: *Barriers, Farming, Prospects, Organic, Constraints.*

INTRODUCTION

India is the home to 30 per cent of the total organic producers in the world, but accounts for just 2.59 per cent of the total organic cultivation area of 57.8 million ha, according to the World of Organic Agriculture 2018 report. At the same time, most organic farmers are struggling due to poor policy measures, rising input costs and limited market as reported by the Associated Chambers of Commerce and Industry of India (ASSOCHAM). The key issues that hinder adoption of organic farming in India are: reduced yield, non-availability of sufficient organic supplements, bio-fertilizers, integration of livestock, certification constraints, high input cost, ecology, marketing and policy support. Capital driven regulation by contracting firms further discourages small farm holders in adopting organic farming. It has been argued that organic farming is productive and sustainable when practiced scientifically. Further, an integrated effort is needed from government and NGOs in the form of subsidies, agriculture extension services and research, to adopt organic farming.

Organic and Development of Organic Farming in India

India has traditionally practiced organic agriculture, but the advent of green revolution technologies led to the increased use of chemicals. In recent years, however, limitation of agriculture based on chemical use and intensive irrigation has become apparent. There has been a resurgence of interest in organic agriculture mainly due to two concerns, decline in crop yield in certain areas as a result of inter alia excessive use of chemical inputs, decreased soil fertility, environmental awareness and health awareness. Also, it is generally perceived that the organically grown produce is environmentally safer and more nutritious than conventionally grown produce. Intensive farming exacerbates phenomena like soil erosion, loss of organic matter (Montgomery, 2007 and Lal, 2010), considered to be great threat for the future sustainability of crop production on a global scale, especially under extreme climatic events such as droughts. The dual advantage of organic farming being the positive effects on the environment as well as to make the farmer self-sufficient as far as his requirements regarding agro-inputs is concerned (Prasad and Gill, 2009) that too at a reduced price or attractive source of rural income generation (Reddy, 2010 and Shukla *et al.*, 2013).

The 10th five-year plan encouraged the promotion of organic farming using wastes, and Integrated Pest Management (IPM) and Integrated Nutrient Management (INM) practiced. Even the 9th five-year plan had emphasized the promotion of sustainable agriculture. The government of India has also launched the National Programme for Organic Production (NPOP) in the year 2001. As per the available statistics, India occupies 9th rank in terms of world's Organic Agriculture land and stands 1st in terms of total number of producers as per 2018 data of FIBL and IFOAM. According to the Agricultural and Processed Food Product Export Development Authority (APEDA), as on 31st March 2018, total area under organic certification process (registered under National Programme for Organic Production) is 3.56 million ha (2017-18). This includes 1.78 million ha (50 per cent) cultivable area and another 1.78 million ha (50 per cent) for wild harvest collection. Among all the states, Madhya Pradesh has covered largest area under organic certification followed by Rajasthan, Maharashtra and Uttar Pradesh. During 2016, Sikkim has achieved a remarkable distinction of converting its entire cultivable land (more than 76,000 ha) under organic certification.

The states of Uttarakhand and Sikkim have been declared as 'Organic States'. India produced around 1.70 million MT (2017-2018) of certified organic products which includes all varieties of food products namely oil seeds, sugarcane, cereals and millets, cotton, pulses, medicinal plants, tea, fruits, spices, dry fruits, vegetables, coffee *etc.* The production is not limited to the edible sector but also produces organic cotton fibre, functional food products *etc.* Among different states Madhya Pradesh is the largest producer followed by Maharashtra, Karnataka, Uttar Pradesh and Rajasthan. In terms of commodities, oil seeds is the single largest category followed by sugar crops, cereals and millets, fibre crops, pulses, medicinal, herbal and aromatic plants and spices and condiments.

Prospects of Organic Farming

Consumer Acceptance

Consumers are now turning to organic food because they believe it is tastier, as well as healthier, both for themselves and environment due to awareness of environmental degradation related issues. The emergence of organic shops in market creates the awareness to stimulate the customer need and demand. The concern regarding to organic agriculture rise now a days is essential in preserving the environment and improving public health. Another reason for organic products prominence is the opposition to genetically modified food. The rainfed areas particularly north-eastern regions where least or no utilization of chemical inputs due to poor resources provides considerable opportunity for promotion of organic farming thereby reflecting its vast but unexplored scope. It has also been reported that organic food consumption is not only related to health aspects or taste of the product itself but often to values such as altruism, ecology, spirituality and sometimes combined with an alternative life style. Most of the consumers are unfamiliar with organic farming and processing standards but they are aware of its central features such as "chemical free", "café free" or "natural" and see therefore a "plus" in food quality, a reason to purchase organic foods.

Environment Friendly

Pest-Control

Organic farming is environment- friendly. Because it is well know that use of chemicals have led towards destruction of many beneficial insect species and have caused environmental degradation. For instance, Korean researchers have reported that avoiding pesticides in paddy fields encourages the muddy loach fish, which effectively controlled mosquitoes that spread malaria and Japanese encephalitis.

Higher Biodiversity

Organic farming also provides energy for microbial activity and this has been suggested as an indicator of microbial biodiversity. It is an indicator of change in soil properties, as the size and activity of the microbial quotient is directly related to the amount and quality of carbon available. Organic farms, often explore more biodiversity than conventional farms because it usually has more trees, a wider diversity of crops and many different natural predators, which control pests and help prevent disease.

Better Soil Health

Increasing soil organic matter by organic farming has the added benefit of improving soil quality and thereby enhancing the long-term sustainability of agriculture. Increases in crop yields during the initial years of organic farming have been attributed to gradual improvements in soil properties, such as the capacity of the soil microbial community to mineralize nitrogen. Thus, organic production systems have the potential to achieve sustainability of agricultural system.

Problems in Adopting Organic Farming

Yield Reduction in Conversion to Organic Farms

Yield in irrigated farm usually go down during the conversion period from conventional to organic farming. The crop yields are boosted by artificial fertilizers while it takes time for the soil fertility to get boosted through organic farming. However, after conversion, yields are gradually improved and become many a time equal, if not higher that under the conventional farming after 4-5 years of practicing organic agriculture. Problems are evident even in Sikkim, which has been recognized as the country's first organic state in 2018. A survey by Delhi-based Centre for Science and Environment shows that the state's transition to organic farming is yet to become a true success. With the use of organic sources of nutrients, the yield of the crop is very low (Kaur and Toor, 2015), especially during initial stages, although it may stabilize later, yet complete dependence on pure organic farming will not be sustainable in the long run. The survey found that the phasing out of chemicals in Sikkim was not complemented by a simultaneous increase in availability of and access to organic manure. Farmers also complained of low productivity during the transition from conventional chemical farming to organic farming. Pest attack on organic crops is another reason cited by the farmers for low productivity and demanded education and training to deal with it. Similarly, nearly 98 per cent farmers in Rajasthan are aware of ecological hazards of conventional chemical-based farming, but fear of decline in production and unavailability of organic inputs in the market discourage them from switching to organic farming, says a 2015 study conducted by the Consumer Unity and Trust Society.

Increasing Pest and Weeds

One of the foremost challenges is the rampant use of pesticides and chemicals for weeds under conventional farming. This has caused an evolution of the resistant pest and weeds species and thereby, made them increasingly immune to chemicals. This is the first hurdle in the transition from conventional farming to organic farming. As organic farming prohibits synthetic pesticides, the vulnerable farmer is at the mercy of severe attack from mutant pest. The situation becomes worse not only with the discovery of newer variants of pest traditional methods of pest control that fail to prevent the damage to crops.

Resource Need

Livestock resources have played multifaceted roles strengthening indigenous agricultural practices and generating income and livelihoods for farmers in India. However, with the advent of technologies in agriculture, livestock population consisting of cattle, buffalo, sheep, goat, pig, horses and ponies, mules, donkeys, camels, mithun and yak in the country is 512.05 million in 2012. The total livestock population has decreased by about 3.33 per cent over the previous census. The decline for cattle, sheep, goats and pigs has been 4.10, 9.07, 3.82 and 7.54 per cent respectively. Advent of technology in agriculture coupled with inadequate feed supply is the major determinant reducing attraction of common people towards livestock production and thus availability of manure, even to small farm holders. Nature pastures, crop residues and kitchen wastes are the major feed components in India. The area under natural pasture in India is rapidly declining as a result of expanding urban and agricultural areas. Since, organic supplements are required in bulk, their production, collection, transportation, storage and application may further be impeded by constrained labour and transport facilities for areas away from the site of production/collection. A major part of rural population in India is poor and use animal manure for soil amendment. In a typical rural household in India, about 1,5000-2,000 kg (dry weight basis) of cattle manure is used as domestic fuel annually. Also, most of crop residues are removed from the fields for use as fodder and fuel. A decrease in area under natural pasture creates tremendous pressure on crop residues to be used as fodder. This together with rising use of crop residues as in rural sectors has resulted in rare use of residue as mulch.

Certification

The cost involved in the certification and a time lag of three years (conversion stage) is the major constraint for small land holder farmers in India, for adopting organic farming. Organic produce needs certification to ensure that all synthetic inputs are prohibited and soil building approaches are followed. The certification process aims at converting the growing area to comply with requirement of standards within a period of 3 years, for this reason farmer who want to adopt organic management become bound to wait for minimum of 3 years under certification procedure which is essential for purging of chemical residues. The certification systems are not only cumbersome and time-consuming, but also expensive. It is important to eliminate confusion over multiple ministries regulating organic production and sales. In fact, FSSAI is reconsidering its regulations on organic food that kicked in from July 1, 2018, after farmer's protests. The government, meanwhile, has not done enough to address the hurdles. Paramparagat Krishi Vikas Yojana (PKVY)-the Centre's free certification programme for organic farmers is an example. A 2018 report on the implementation of PKVY highlights that all states, except Tripura, Odisha and Karnataka, have failed to utilize even 50 per cent of their funds under the scheme. While the Centre has increased allocation for the scheme by 44 per cent for the current year, corrective measures are needed to ensure that the states become responsible and contribute toward organic India.

Technical Know-how

Farmers also face a number of constraints such as lack of technical know-how, for example organic farming practices and production methods, and lack of market information, which market and distribution channel to choose, which commodity/products to produce, competition, market access *etc.*

Lack of Clarity

Organic farming is still hampered by lack of clarity. Consumers were not always sure about what was really covered by organic farming and the restriction it implied.

Marketing

The domestic market for organics is undeveloped in India. Lack of domestic marketing channels adds to the difficulties faced by the farmers converting to organic methods in agriculture. It had been found that the farmers of organic wheat in Rajasthan got lower prices than those of the traditional wheat (Rao, 2003).

Policy Support

Policies have long been focused on generating external solutions to farmer's needs. It has encouraged dependencies on external inputs, though they are more costly, environmentally inefficient when compared to the resource-conserving options. Despite serious efforts of some NGOs it appears that India is lagging far behind in the adoption of organic farming. For laying the spade work for the country, certain issues require attention at government policy making levels.

Lack of Infrastructure and Cold Storage Facility

The difficulties of an organic farmer do not end with harvesting of produce. Serve lack of suitable infrastructure is proving to be the weakness for organic farming and its produce. Maximum organic farming is practiced for fruits and vegetable in nature. The challenges posed by inadequate agriculture infrastructure and cold storage facilities translate to loss of produce due to spoilage. Additionally, poor road infrastructure especially in the hilly states, result in poor and delayed connectivity to farmer markets. Once the farm is established organic, the yield increases and therefore the cost of production decline (Pandey and Singh, 2012).

Organic Farming Fraught with Hurdles

According to the Indian Council of Agricultural Research, productivity on an average dips by 6.7 per cent in the first year, and the government needs to have a plan in place to support farmers during the transition. The report on Doubling of farmer's Income by Ashok Dalwai committee, too, echoes the concern of the farmers who claim up to 30 per cent drop in yields when embracing organic. It takes about a decade to attain pre-conversion yield levels, according to the committee report. To ensure marketing of organic produce, connecting farmers with the domestic and global supply chain is extremely important. But a new ICRIER study found the supply chain to be underdeveloped. As a result, small and mid-size farmers located in hilly regions and tribal belts find it extremely difficult to access the market.

Expensive Organic Produce Discourages Customers, Affects Sales

Even after the produce reaches the markets, consumers find them expensive and discouraging. Associated Chambers of Commerce and Industry of India (ASSOCHAM) said that ₹1200-1500/month is the additional expenditure if a consumer switches to organic food. Farmers said that their organic produce are usually more expensive due to higher labour cost and comparatively lower yields. According to the ASSOCHAM report, postharvest handling of relatively small quantities of organic foods also results in higher costs because of the mandatory segregation of organic and conventional produce, especially for processing and transportation. Specialized farmers training costs, processing and inventory holding costs (without chemical additives), and increased packaging, logistics and distribution costs (due to low volumes), contribute to the high price of organic food products.

CONCLUSION

Research in organic farming has increased considerably in the developing countries recently, but there is still a need for targeted research activities to help promote and develop organic farming. Presently there is lack of government subsidies or support to make conversion to organic status easier and cheaper. The question about yield and financial viability of organic farming are crucial and there are no empirical studies available in the Indian context comparing the economic and ecological returns of organic farm vis-à-vis conventional farms. Organic agriculture has been neglected in the agriculture policy, and therefore there is less government assistance for the promotion of organic agriculture in the form of subsidies, agricultural extension services and official research.

REFERENCES

Gurav, P. P., Chaudhari, P. L., Lakaria, B. L., Jha, P. and Shinogi, K. C. 2019. Constraints in organic farming. *Indian Farming*69(08): 16-19.

Kaur, K and Toor, M.S. (2015). Organic Farming: Status and Constraints. *Indian J. Econ.* Dev., **11**(1): 333-338.

Lal, R. 2010. Enhancing eco-efficiency in agro-ecosystems through soil carbon sequestration. *Crop Science* **50**: 120– 131

Montgomery, D. R. 2007. Dirt: The Erosion of Civilization. University of California Press, Berkeley.

Pandey, J. and Singh, A. 2012. Opportunities and constraints in organic farming: an Indian perspective. *Journal of Scientific Research* **56**: 47-72.

Rao, P. S. 2003. Marketing of organic produce (wheat), in Rajasthan district of Rajasthan. *Indian Journal of Agriculture Marketing,* conference number special.

Reddy, B. S. 2010. Organic farming: status, issues and prospects – a review. Agric. Econ. Res. Revision, 23: 343-358

Shukla, U. N., Manju, L. M. and K. C. Bairwa. 2013. Organic Farming: Current Status in India Popular Kheti, (**1**) 4(October-December)

Wani, S. A., Wani, M. A., Meraj, S., Padder, A. B. and Chand, S. 2017. Organic Farming: Present status, scope and prospects in northern India. *Journal of Applied Natural Science* **9**(4): 2272-227.

Chapter 4

Remote Sensing: A Tool in Agriculture

***Amitesh Kumar Singh*[1]*, Ravi Shanker*[2] *and Amit Kumar Singh*[2]**

[1]*Krishi Vigyan Kendra, Kallipur, Varanasi (ANDUAT, Ayodhya), Uttar Pradesh*
[2]*Gramin Vikas Trust, Krishi Vigyan Kendra, Chakeshwari Farm, Godda–814 133, Jharkhand*

ABSTRACT

Agriculturist has researched and used remote sensing technology for acreage estimation since the early 1970s. Significant advancements in recent years have enabled them to transition the use of remote sensing from primarily a research function to performing an integral role in the agency's crop acreage estimation program covering all major crops grown in high producing states. This accomplishment was achieved in large part as a result of enhanced data partnerships, improved methodologies, increased availability of commercial software, and improved imagery and ancillary data. Although, yield estimates are solid, with more experience and research, improvements are expected. Looking to the future tools of research to quantitatively measure crop progress and condition using remote sensing and intends to expand the use of remote sensing into mapping soil moisture and improving disaster assessments and monitoring.

Keywords: *Biomass, Tools, Remote sensor, Agriculture, Phenological.*

INTRODUCTION

Remote sensing techniques are widely used in agriculture. The use of remote sensing is necessary, as the monitoring of agricultural activities faces special problems not common to other economic sectors. First of all, agricultural production follows strong seasonal patterns related to the biological lifecycle of crops. The production depends secondly on the physical landscape (*e.g.*, soil type), as well as climatic driving variables and agricultural management practices. All variables are highly variable in space and time. Moreover, as productivity can change within short time periods, due to unfavorable growing conditions, agricultural monitoring systems need to be timely. This is even more important, as many items are perishable. Thus, as pointed out by the Food and Agriculture Organization (FAO) the need for timeliness is a major factor underlying agricultural statistics and associated monitoring systems information is worth little. Remote sensing can significantly contribute to providing a timely and accurate picture of the agricultural sector, as it is very suitable for gathering information over large areas with high revisit frequency. The present paper summarizes the main remote sensing applications, with a focus on regional to global applications. It provides arguments for enhancing investments in agricultural monitoring systems. It follows the strong conviction that a close monitoring of agricultural production systems is necessary, as agriculture must strongly increase its production for feeding the nine-billion people predicted by mid-century. This increase in production must be achieved while minimizing the environmental impact of agriculture. Achieving this goal is difficult, as agriculture must cope with climate change and compete with land users not involved in food production (*e.g.*, biofuel production, urban expansion, *etc.*). The necessary changes and transitions have to be monitored closely to provide decision makers with feedback on their policies and investments. Together with the emergence of a range of commercial GIS packages and other software tools for the manipulation of spatially referenced datasets. This has facilitated the emergence of a range of new application of satellite data which have been developed or have entered operational service over the past decade.

Although there have been many demonstration of the value of satellite to development issues such as food production, resources management and environment characterization, it is recognized that there is a role for transferring that knowledge to developing countries, and it is hoped that this initiative encourages and assist countries inexperienced in the use of satellite data to acquire the necessary expertise to exploit and realize its full potential. The raw data from satellite often requires complex processing both to correct for atmospheric and geometric distortions and to derive information from the data and imagery. Additional data, such as that from *in-situ* sources, is sometimes used to supplement the data in this processing. Such data is particularly useful for calibrating or validating models.

This review aims to provide an overview of recent remote sensing developments in terms of regional and global applications for agriculture. To illustrate the urgent need for an enhanced monitoring capacity, the paper is structured as follows. It includes methods involving different types of satellites used along with statistical methods. It involves types of remote sensing *i.e.* active and passive. It is concluded by the fact that developments in computer technology over the past few years have resulted in the availability, at relatively low cost of compact, high performance computer, which are well suited to the demands of satellite data processing.

1. Remote Sensing of Agricultural Resources

Already in the early 80s, it was shown by Tucker and co-workers that green vegetation can be monitored through its spectral reflectance properties. Today, a large range of satellite sensors provide us regularly with data covering a wide spectral range (from optical through microwave) Lee 2000. Data are acquired from various orbits and in different spatial and temporal resolutions. For deriving the sought information, a large number of spectral analysis tools have been developed. Besides the spectral signature, useful information can also be retrieved by analyzing the temporal signature and directional reflectance properties of vegetation. Further useful information can also be retrieved from the spatial arrangement of the pixels, *i.e.*, the texture of the image, even at coarse resolution.

Biomass and Yield

In the early 80s, Tucker and co-workers found that arithmetic combinations of vegetation reflectance's in the red and near infrared (so called "vegetation indices" or VI) are particularly useful for vegetation characterization. For example, the well-known NDVI (Normalized Difference Vegetation Index) was already proposed in 1978 by Deering. The index became, subsequently, the most popular indicator for studying vegetation health and crop production. The success of the NDVI stems from its close relation to the canopy Leaf Area Index (LAI) and fAPAR (fraction of Absorbed Photosynthetically Active Radiation). Due to its almost linear relation with fAPAR, the NDVI can be readily used as an indirect measure of primary productivity. Rembold *et al.* (2013), a comprehensive overview is provided regarding biomass and yield mapping approaches. Different methods are discussed, ranging from simple regression equations, to the use of more complex crop growth models. For this reason, the present section focuses only on the so-called 'yield correlation masking'. Strong reference is made to the excellent publication of Kastens *et al.*, 2005. It has also to be noted that time integration analysis generally increases the prediction accuracy.

One obstacle to successful modeling and prediction of crop yields using remotely sensed imagery is the identification of image masks. Image masking involves restricting an analysis to a subset of a region's pixels, rather than using all of the pixels in the scene. Cropland masking, where all sufficiently cropped pixels are included in the mask regardless of crop type, has been shown to generally improve crop yield forecasting ability.

Vegetation Vigor and (Drought) Stress Monitoring

In agricultural monitoring, timely and specialized information about general growing conditions of the vegetated land is of high importance. For this reason, the use of satellite imagery for plant stress monitoring became an important remote sensing application. The three individual drought indices PDI (Precipitation Drought Index), TDI (Temperature Drought Index) and VDI (Vegetation Drought Index—as a substitute for the Soil Moisture Drought Index)—are combined as a weighted sum to yield the CDI (Equation 1):

$$CDIi,m= wPDI^*PDIi,m+wTDI^*TDIi,m+wVDI^*VDIi,m$$

By definition, CDI = 1.0 represents average weather conditions. If the CDI is greater than 1.0, it represents wetter than average; if it is below 1.0, it represents dryer than average conditions. Drought severity categories may be fitted to geographic locations and/or plant types.

Crop Phenological Development

The phenological dynamics of terrestrial ecosystems reflect the response of the Earth's biosphere to inter- and intra-annual dynamics of the Earth's climate and hydrologic regimes. Remotely sensed satellite data possess significant potential for monitoring vegetation dynamics, due to their synoptic coverage and frequent temporal sampling. This enables the monitoring of simple phenological events, such as the start and peak of vegetation growth, both in natural ecosystems and in agricultural landscapes.

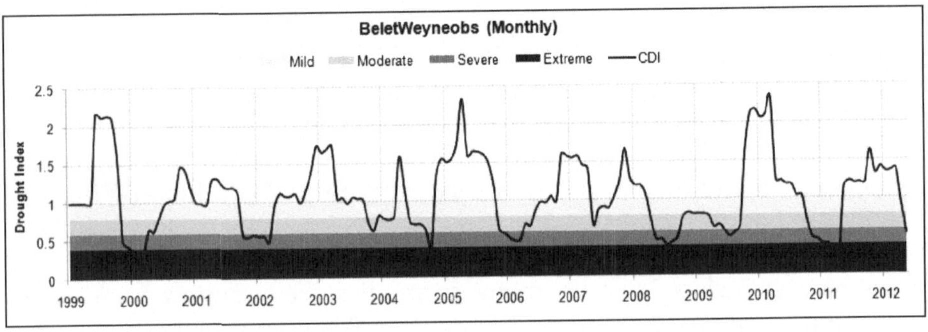

Figure 4.1: Time Profile of CDI (Combined Drought Index).

Crop Acreage Estimation and Cropland Mapping

Cropland areas are often characterized by a diverse mosaic of LULC types that change over various spatial and temporal scales in response to different management practices and agricultural policies. As a result, detailed regional-scale cropping patterns need to be mapped on a repetitive basis. Such information is also necessary to better understand the role and response of regional cropping practices in relation to various environmental issues (*e.g.*, climate change, groundwater depletion, soil erosion) that potentially threaten the long-term sustainability of major agricultural producing areas. Of course, for each growing season, the acreage of the different crops must be known for accurate production estimates. The excellent work of Wardlow *et al.* (2007) provides a valuable discussion of different approaches.

Disturbances and Land Use/Land Cover Changes

During different stages of vegetation growth, plant structures and associated pigment assemblages can vary significantly Mueller *et al.* (2009) Thus, the same vegetation type can appear significantly different and different types similar, at various stages during intra-annual growth cycles. A fine-tuning of change thresholds is thus necessary to separate 'real' changes from 'natural' variations.

2. Critical Component for Remote Sensing

The first critical component for any remote sensing program is solid ground truth information. Without ground data to identify land cover categories, to train the classifier and validate the output image products, it is impossible to run a defensible program that provides reliable results Johnson 1987. Ground truth is mentioned first, because it must be seriously considered before initiating plans for any remote sensing application. Secondly, a source of satellite imagery is required.

Ground Truth: Imagery

Characteristic Features

1. Synoptic Nature
2. Repetitive Coverage of Large Area
3. ***Quantifiable Data Procurement*** REGIONS OF ELECTROMAGNETIC SPECTRUM

Requirements

Platform: SATELLITE (geostationary and polar)

Instrument: SCANNER

Model: Example

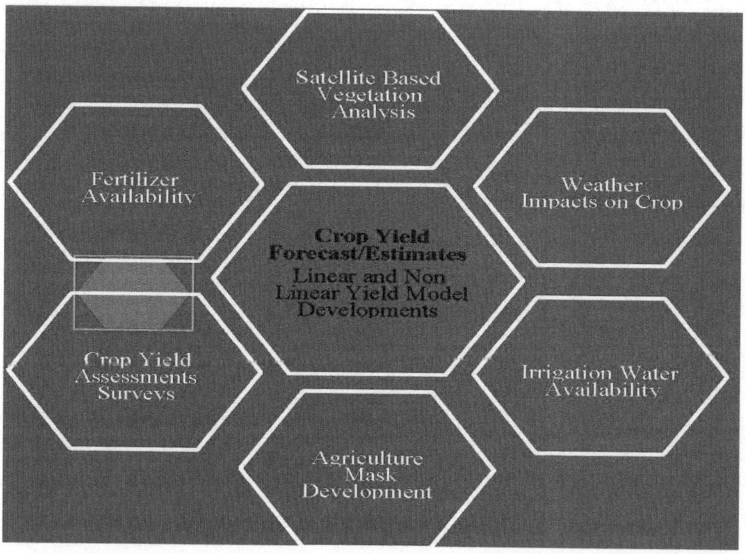

CONCLUSION

The review demonstrates the strong role remote sensing plays within the agricultural sector. The remotely provided information is urgently needed for various decision makers. Requests for objective information will increase in the future, as a result of the expected changes in the agricultural sector (*e.g.*, meeting food requirements and environmental restrictions) Maselli (1993). For example, for most crops, large production increases (between 45 and 70 per cent) are possible from closing yield gaps to 100 per cent of attainable yields. Remotely sensed information can help with identifying yield gaps and monitoring related agricultural practices. Using 'yield-correlation masking', this would even be possible without detailed crop masks, provided enough (historical) yield data is available. In parallel, environmentally sensitive areas could be identified for protective purposes. The detection of phenological indicators not only helps with identifying different land covers and crops (and crop growing conditions), but at the same time, provides evidence for ongoing global/climate change. With appropriate pre-processing of time series (gap filling and smoothing), phenological indicators, such as start of the growing season, can probably be estimated with accuracies of ±7 to 10 days. Vegetation anomalies related to local meteorological conditions (*e.g.*, droughts) can be readily detected from space and combined with other data sources to indicate stress affected regions. This information is not only important for organizations dealing with food security, but can also help with identifying a region's vulnerability to (drought) stress. Finally, the detection and monitoring of (permanent) land-cover changes is, for example, important for establishing links between policy decisions, regulatory actions and subsequent land-use activities.

REFERENCES

Foody, G.M. 2010. Assessing the accuracy of land cover change with imperfect ground reference data. *Remote Sens. Environ., 114,* 2271–2285.

Johnson, G.E.; van Dijk, A.; Sakamoto, C.M. 1987. The use of AVHRR data in operational agricultural assessment in Africa. *Geocarto Int., 1,* 41–60.

Kastens, J.H.; Kastens, T.L.; Kastens, D.L.A.; Price, K.P.; Martinko, E.A.; Lee, R.-Y. 2005. Image masking for crop yield forecasting using AVHRR NDVI time series imagery. *Remote Sens. Environ., 99,* 341–356.

Lee, R.; Kastens, D.L.; Price, K.P.; Martinko, E.A. 2000. Forecasting Corn Yield in Iowa Using Remotely Sensed Data and Vegetation Phenology Information. In *Proceedings of the Second International Conference on Geospatial Information in Agriculture and Forestry,* Lake Buena Vista, FL, USA, 10–12 January; Volume II, pp. 460–467.

Maselli, F.; Conese, C.; Petkov, L.; Gilabert, M.A. 1993. Environmental monitoring and crop forecasting in the Sahel through the use of NOAA NDVI data. A case study: Niger 1986–89. *Int. J. Remote Sens., 14,* 3471–3487.

Mueller, R., Boryan, C. and Seffrin, R. 2009. Data Partner Synergy: The Cropland Data Layer. August '09 GMU conf http: //www.geoinformatics2009.org/

Rembold, F.; Atzberger, C.; Rojas, O.; Savin, I. 2013. Using low resolution satellite imagery for yield prediction and yield anomaly detection. *Remote Sens.*, in review.

Wardlow, B.D.; Egbert, S.L.; Kastens, J.H. 2007. Analysis of time-series MODIS 250 m vegetation index data for crop classification in the US Central Great Plains. *Remote Sens. Environ.*, *108*, 290–310.

Chapter 5

Brown Manuring: Tools for Nutrients Management

R.P. Singh[1], S.K. Bairagi[2], A.P. Singh[3] and Raksha[4]*

[1]*Department of Agronomy,* [2]*Department of Horticulture,* [3]*Department of Ag. Chemistry, Amar Singh College, Lakhaoti, Bulandshahar, Uttar Pradesh*
[4]*Department of Home Science, Siddharth University, Kapilvastu, Siddharth Nagar, Uttar Pradesh*
**e-mail: raj1977887@gmail.com*

ABSTRACT

The efficient nutrients management is required to every crop to achieve suitable crop yield. Organic manure can be play in important role in achieving sustainability. But these manure are bulky in nature, and low in nutrients content, hence replacement is highly required. Brown manuring other possible and best options of providing nutrients to the crop from organic sources in short period 20-30 days in transplanted rice and 50-60 days in direct seeded rice and wheat brown manuring also advantageous for weed management, increase moisture holding capacity, organic carbon (per cent) and yield of crops. In dryland area of India, scarcity of water is very common so it is not possible to grow green manuring crop before rainy seasons in this situation brown manuring is the best options. Brown manuring is a best tool for nutrients management.

Keywords: *Brown manuring, Nutrients management, Organic carbon.*

INTRODUCTION

The present agriculture scenario is concerned with rising cost of cultivation and increased the agriculture inputs. Among the agriculture inputs chemical fertilizers is the major input required for production of crops. The chemical fertilizers made a major contribution to the remarkable increase in crop yield.

The continuous application of chemical fertilizers degraded the quality of produce as well as their harmful effect on soil, plant, animals and micro-organism health. It is well known that plants required 17 essential elements to complete their healthy life cycle but now day's farmers of India basically apply primary nutrients (NPK). Therefore, increases imbalance in other nutrients supplies. The combined use of organic manure, cultural practices and Biofertilizers help to preserve yield stability by reducing marginal secondary and micronutrients shortages, improving the productivity of applied nutrients and maintain favorable physical conditions in the soils. India has sufficient bio-waste potential that can be converted into compost and organic manures and used in crop field. But the enormous amount of organic waste is not being properly scientifically transformed into composts and lost by various means. Besides bulky organic manures, use to sustain soil fertility.

Concept of Brown Manuring

Brown manuring was first introduced in North -south Wales, Australia, lock heart district in 1996. Goodness was the first person to adopt the practice of Brown manuring. In Australia for winter crop they employed this strategy against the herbicide-resistant ryegrass population. They planted a cover crop with the intent to incorporate it into soil before weed seed setting. This practice has help to rotate chemical classes, preserve ground cover, prevent weed seed setting and add useful nitrogen from pulse nitrogen fixation, all well as provide the farm cropping system with agronomic advantages of improve soil quality and water holding capacity.

Brown manuring may be defined as growing of green manuring crops like sunhemp, moong, cowpea *etc.* as a inter or mixed crop (basically legume crops) and killing them which attain the age of 24 days, Applying selected herbicide like 2,4-D sodium salt 80 percent w.p.@ 0.75 kg per hectare. After 4-5 days of spraying green manuring crop plant (Sesbania, Sunhemp *etc.*) appear brown and then start drying as it is selective herbicide that kill only green manuring crop and not the cereals crops (rice, wheat *etc.*). This is called knocking down effect. Brown manuring is usually recommended for the monocot crop the rice is transplanted or sown in the field both condition is suitable.

Sesbania with Systems of Rice Intensification (SRI)

A system of rice intensification (SRI) methods is the new methods of rice transplanting. In which the rice seedling is transplanted 25cm x 25cm at the age of 12 DAS, alternate wetting and drying condition is also provide to the field at time to time and cono-weeder is use for pulverized the soil, mixed weeds into the soil. After 2-3 days transplanting of paddy the Sesbania seeds broadcast @35 kg per hectare. Sesbania germinate and attain the age of 24 days (Biomass at list 1kg per hectare) the selective herbicide like 2,4-D sodium salt 80 per cent w.p. @ 0.75 kg per hectare applied to kill sesbsnia plants. After 5-6 days application of herbicide sesbania plants dry and look like brown in colour this reason methods known as brown manuring. After browning of sesbania plants light irrigation given to the crop than cono-weeder is used to mixed in to the soil. This method pulverized the soil, control the weeds, increase aeration, carbon content and water holding capacity of soil ultimately maximized yield of the crop.

Deference between Green Manuring and Brown Manuring

Green Manuring	*Brown Manuring*
For cultivation of green manuring required tillage practices.	It is no-till version of green manuring, where herbicide to kill the green manure crops and weeds.
Crop of green manuring incorporate into the soil before flowering.	Crop of brown manuring kill by herbicide at the time of 24 days after sowing.
Soil erosion problems.	There is no risk of soil erosion.
Soil moisture is necessary for incorporation and decomposition of green manuring crop.	Moisture is not necessary but use of cono-weeder moisture is required.
The microbial population is necessary for decomposition.	Chemical herbicide required to kill the green manuring crops.

Crops Used for Brown Manuring

The leguminous crops used for green manuring and brown manuring both because these crop having less C: N ratio, succulent plant type, air nitrogen fixed into root nodules and easily decomposed by living micro-organism. The non leguminous crop used as a green manuring crop only increase carbon content into the soil and also late decomposed by micro-organism. The major leguminous crops used as brown and green manuring as follows: Sunhemp, green gram, Dhaincha, cowpea, lentil *etc.*

Qualities of an Ideal Brown Manuring Crop

Crop species that are most suitable to Brown manuring enable growers to maximize weed control and nitrogen fixation while minimize cost and risk.

The main criteria to be considered in selection of brown manuring crops include:

1. Seeds of the plants should be easily availability and cost effective
2. It should be easy to cultivate and have vigor growth
3. Green manuring crop have capacity to produced more dry matter per unit area of land
4. It should have competitiveness with all type weeds

5. The crops should have high ground cover to reduced water and wind erosion
6. The crop should have capacity to well grow under unfavorable soil condition

Advantages of Brown Manuring

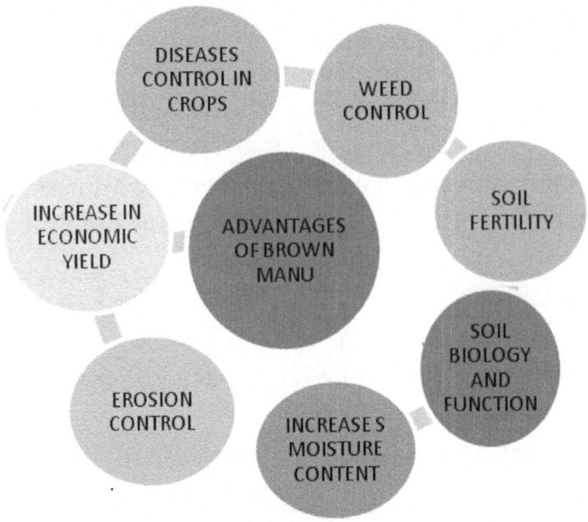

This figure indicate that brown manuring increases soil organic matter content and also supply the required amount of nitrogen to the crop.

- The biomass of the green manuring crop increase moisture holding capacity of the soil.
- It also increase parameters like organic carbon content, nitrogen content and provide suitable environment for growth of earthworms.
- Brown manuring improve the physical, biological and chemical properties of soil.
- Brown manuring reduced the weeds population due to its fast growing habits.
- It increase the organic matter of soil and decrease the bulk density of soil which act as buffering substances

Effect of Brown Manuring on Rice Agro-ecosystems

Brown manuring affect soil properties and all the phases of crop growth like seed germination to maturity. There is very little works have been done in the field of brown manuring. The article emphasis to address potential of brown manuring on soil health to be obtain maximum yield. In rained area of India having huge scope to increase soil fertility.

Table 5.1: Effect of Brown Manuring on Soil Organic Carbon and Postharvest Nitrogen

Year	Initial Carbon Content of Soil (Per cent)	Organic Carbon Content after Harvest (Per cent)	Percent of Organic Carbon Increase	Initial Soil Nitrogen Content (kg/ha.)	Soil Nitrogen Content after Harvest (kg/ha.)	Per cent of Increase in Soil available Nitrogen
2015	0.53	0.68	0.15	282.10	318.21	36.26
2016	0.59	0.70	0.11	284.84	323.64	38.80
Mean	0.56	0.69	0.13	283.47	320.67	37.48

The experiment was conducted in GVT-Kirshi Vigyan Kendra, Godda, Jharkhand farm under on station trail 2015-16. It is depicted from the data (Table 5.1) that both the parameters increase under brown manuring rice agro-ecosystems. Similar finding was also given by Sarangi *et al.*, 2016. Growing of direct seeded rice with sesbania crop increase the organic matter and nitrogen content of soil.

CONCLUSION

Brown manuring is a technique when plant material return back to the Agro-ecosystems to recover soil fertility and increase physical, biological and chemical structure of soil. It also conserve the soil water reduced the weeds and disease infestation problems along with counterbalance of green house gas emissions. In rained and rice growing area these are best technology to maintain soil fertility. Continuous application of chemical fertilizers creates several problems like soil, water, environment pollutions and also hampered health of ecosystems. As the brown manuring practice is eco-friendly and improve over all ecosystems health. These are new technique need to spread in farming community by different agencies.

REFERENCES

Edward T. 2020. Carbon farming: green and brown manuring. Agriculture and food Retrieved from: https: //www.agric.wa.gov.au/carbon-farming-green and brown manuring.

Sarangi D. R., Sahoo, T. R., Sethy, S., Chaurasia, M., Prasad, S. M. and Mohanta, R. K. 2016. Effect of replacing a part of nitrogenous fertilizers by brown manuring in direct seeded rice. *Oryza* 53(3): 226-228.

Sharma, A. R. and Ghosh, A. 2000. Effect of green manuring with Sesbania aculeate and nitrogen fertilizers on the performance of direct seeded rice flood prone lowland rice Nutrient cycling in agro-ecosystems. 57: 141-153.

Chapter 6

Nutrient Management in Field and Horticultural Crops

Ravi[1]*, Mukesh Kumar[2] and Piryanshu[3]

[1]Assistant Professor, Department of Soil Science and Agricultural Chemistry,
[3]Assistant Professor, Department of Horticulture,
R.S.M. (PG) College Dhampur (Bijnor)–246761, Uttar Pradesh
[2]Assistant Professor (Soil Science), Department of Natural Resource Management, College of Horticulture, S.D. Agricultural University, Jagudan–384 460, Gujarat
**e-mail: ravi.slusi@gmail.com*

ABSTRACT

Organic farming, agricultural system that uses ecologically based pest controls and biological fertilizers derived largely from animal and plant wastes and nitrogen fixing legume crops. This system facilitates the build-up of soil organic matter, reducing risk of erosion and runoff and enhancing nutrient storehouse in soils for plants. FYM, compost and green manuring are the most important and widely used bulky organic manures. Manuring with different short-duration legumes is suitable for maintenance of soil quality in terms of adding nitrogen to soil. Sustainable quantity of essential plant nutrients can be maintained by the use of biomass available in the form of leaf fall or plant residues. Manures, legumes, crop residues and bio-fertilizers are the main ingredients of organic nutrient management. India produces about 1800 MT of animal dung per annum. Even if ⅔ of the dung is used for biogas generation, it is expected to yield about 440 m t/annum of manure, which is equivalent to 2.90 MT N, 2.75 MT P_2O_5 and 1.89 MT K_2O.Organic nutrient management is based on four principles of soil;1)Principle of health, 2) Principle of ecology, 3) Principle of fairness and 4) Principle of care. The aim of this chapter to explain about the organic sources of nutrients and how to utilize them for balanced plant nutrition under organic farming.

Keywords: *Manures, Green manuring, Crop residue, Crop rotation and Bio-fertilizers.*

Introduction

Nutrition in plants may thus be defined as a process of synthesis of food, its breakdown and utilization for various functions in the body. Green plants synthesize their food materials from simple constituents (elements). Therefore, plants require essential (17) and beneficial (4) elements which is taken up either from air and water or soil. In soil, the elements are derived from minerals or are mineralized during the biological breakdown of organic matter. The mineral nutrients are taken up by the plants in the form of ions and incorporated into the plant structure or stored in the cell sap for their growth and development. The continuous availability of plant nutrients is essential for the optimum crop production along with quality. However, constant removals of essential nutrients lead to some nutrient deficiency in the agricultural land. Therefore, it is necessary to supply essential plant nutrients from outer sources for obtaining maximum and quality crop production. Depending on soil properties, various materials such as manures, fertilizers and amendments are widely used to supply essential plant nutrients from the outer sources to manage the soil fertility.

In recent years, a concern is being expressed at global level that chemical based energy-intensive modern agriculture is threatening the natural resource system through chemical contamination of soil and water bodies, global warming and ecological degradation. It is well known now that increased use of chemical pesticides and fertilizers have created chain of problems of soil, environment and water degradation. The rural economy is in ruins because of over dependence of outside inputs in agriculture such as seed, fertilizers, pesticides, growth-promoting chemicals, *etc.*

Organic farming is a system of farm production, with an in-built recycling capacity for waste accepted by environmental and health conscious people. Certified organic products are derived from strategic agronomic production and management system that substitutes the use of synthetically compounded fertilizers, pesticides, growth regulators and antibiotics with biological and organic alternatives. Organic farming thus, offers a foresighted sustainable farming system with a viable alternative to conventional approaches to agriculture.

The concepts of organic agriculture were developed in the early 1900s by Sir Albert Howard, Rudolf Steiner, J.I. Rodel and others who believed that the use of animal manures, cover crops, crop rotation and biologically based pest controls resulted in a better farming system.

It is important to constantly work to build a healthy soil that is rich in organic matter and has all the nutrients that the plants required. Several methods *viz.* green manure, addition of manures and bio-fertilizers, *etc.* can be used to build up soil fertility. These organic sources not only add different nutrients to the soil but also help to prevent weeds and increase soil organic matter to feed soil microorganisms. Soil with high organic matter resists soil erosion, holds water better and thus requires less irrigation. Some natural minerals that are needed by the plants to grow and to improve the soil's consistency can also be added. Soil amendments like lime or gypsum are added to adjust the soil's pH balance. Before application of FYM or compost in the field, it is heated and aged for at least two months reaching and maintaining an internal temperature of 60° to 70°C during decomposition to kill unwanted bacteria and weed seeds. A number of organic fertilizers/amendments and bacterial and fungal bio-fertilizers can be used in organic farming depending upon availability and their suitability to crop.

"Organic farming, agricultural system that uses ecologically based pest controls and biological fertilizers derived largely from animal and plant wastes and nitrogen-fixing cover crops."

A successful farming system relies on the management of organic matter to enhance physico-chemical and biological properties of the soil. The effects of soil organic matter are dynamic as it is a source of gradual release of essential plant nutrients; improves soil structure, its drainage, aeration and water holding capacity; improves soil buffering capacity; influence the solubility of minerals and serves as a source of energy for the development of micro-organisms

According to a conservative estimate, around 600 to 700 MT of agricultural waste is available in the country but it is not managed properly (Jnyanendra *et al.*, 2016). We must convert waste into wealth by converting this biomass into energy, nutrient to starved soil and fuel to farmers. India produces about 1800 MT of animal dung per annum. Even if ⅔ of the dung is used for biogas generation, it is expected to yield about 440 m t/annum of manure, which is equivalent to 2.90 MT N, 2.75 MT P_2O_5 and 1.89 MT K_2O.

Nutrient Management

Nutrient Management refers to the maintenance of soil fertility and of plant nutrient supply at an optimum level for sustaining the desired productivity through optimization of the benefits from all possible sources of organic and biological components in an integrated manner.

The major challenge in organic agriculture is the availability of huge quantities of organic inputs for satisfying the farm demand. The use of animal excreta-based manure alone is not sufficient for meeting the nutrient needs of the crops. It is, therefore, necessary to utilize all the sources available on and off farm effectively.

The concept of organic nutrient management is the continuous improvement of soil productivity on long-term basis through appropriate use of organic manures, green manures, bio-fertilizers and other biological derived materials and their scientific management for optimum growth, yield and quality of crops and intensive cropping systems in specific agro-ecological situations.

Concepts

- ✰ Regulated nutrient supply for optimum crop growth and higher productivity
- ✰ Improvement and maintenance of soil fertility
- ✰ Zero adverse impact on agro–ecosystem quality by balanced fertilization of organic manures, bio-inoculants and other organic source of fertilizers

Determinants

- ✰ Nutrient requirement of cropping system as a whole
- ✰ Soil fertility status and special management needs to overcome soil problems, if any
- ✰ Local availability of nutrients resources (organic and biological sources)
- ✰ Economic conditions of farmers and profitability of proposed nutrient management option

- ☆ Social acceptability
- ☆ Ecological considerations
- ☆ Impact on the environment

Principles of Nutrient Management

Nutrients removed by crops must be returned to the soil:

- ☆ Soil physical conditions should be maintained and upgraded
- ☆ Organic carbon levels of soils should be maintained and enhanced
- ☆ Build-up of a biotic stress should be minimal
- ☆ Degradation of land occurring due to soil erosion must be controlled
- ☆ Soil quality with respect to soil acidity, salinity and solidity or toxic elements build-up must be minimized

Goals of Nutrient Management

- ☆ To ensure productive and sustainable agriculture
- ☆ To reduce expenditure on costs of purchased inputs by using farm manure and crop residue, bio-fertilizers, *etc.*
- ☆ To Prevent degradation of the environment and meet the social and economic aspiration of the farmers without harming the natural resource based agricultural production

Components of Nutrient Management

Organic manures, green manuring, legume crops, crop residues and bio-fertilizers are the main source of nutrient management under organic cultivation.

Manures

The manures are organic in nature, plant or animal origin and contain organic matter in large proportion and plant nutrients in small quantities and used to improve soil productivity by correcting soil physical, chemical and biological properties. Organic manures like FYM, compost (prepared from rural and urban waste), vermicompost, oil cakes, *etc.* have large nutrient potential. Compost and FYM have traditionally been the most important manures for maintaining soil fertility and ensuring yield stability. Other potential organic sources of nutrients such as non edible oil cakes and wastes from food processing industry are also there. Moreover, there are several industrial by products and municipal wastes with fair nutrient potential. However, these nutrient carriers have not been properly evaluated to establish their fertilizer equivalents. There is need to integrate these sources depending on their availability in different crops and cropping systems.

Green Manuring

Green manuring is a practice of ploughing or turning undecomposed green plant tissues into the soil for improving the physical conditions as well as fertility of the soil.

Legume Crops

Legumes have a long-standing history of being soil fertility restorers due to their ability to obtain N from the atmosphere in symbiosis with the bacteria *Rhizobium*. Legumes could provide an important ingredient of nutrient management when grown for grain or fodder in a cropping system, or when introduced for green manuring. Legumes grown as green manure forage or grain crops improved the productivity of rice-wheat cropping system and rejuvenated soil fertility (Yadav *et al.*, 2000).

Crop Residues

Crop residues have several competitive uses and may not be always available as nutrient source, yet in the regions like North-West India where mechanical harvesting is practiced, a sizeable quantity of residues is left in the field, which can form a part of nutrient supply. There are large amounts of residues of other crops like potato, sugarcane, vegetables, *etc.*, which are practically wasted in most cases. Although cereal crop residues are valuable cattle feed, these could be used to supplement the fertilizers wherever available in excess of local needs. Stubbles left in the field in traditional harvesting methods range from 0.5 to 1.5t/ha in case of different crops. When mechanical harvesting is done, this amount is much greater. Stubbles of coarse cereals like sorghum, maize, pearl millet, *etc.* which is generally difficult to collect and burnt during land preparation causing significant loss of plant nutrients.

Bio-fertilizers

Bio-fertilizers are the materials containing living or latent cells of agriculturally beneficial microorganisms that play an important role in improving soil fertility and crop productivity due to their capacity to fix atmospheric nitrogen, solubilise/mobilize phosphorus and decompose farm waste resulting in the release of plant nutrients. The extent of benefits from these micro-organisms depends on their number and efficiency, however, which is governed by the soil and environmental factors.

Organically Approved Soil Amendments

Soil amendments are a mineral/organic material having specific characteristics and used to ameliorate the acid or alkali soils by increase or decrease the pH. The maintenance of optimum soil pH is important for crop production. The correct pH is essential for optimal soil biological activity, especially the mineralisation of organic material into plant available nutrients, which is the main source of nitrogen in organically managed soils. The majority of plant nutrients available in soil under near neutral pH while high and low pH limits availability of these nutrients. By addition of approved soil amendments like gypsum or lime that decreased or increased the soil pH.

Nutrient Management in Field Crops

Recycling of Farm Waste

The large quantity of farm waste produced after harvest of field crops like rice and wheat that can be used either as directly incorporated in to soil or after preparation of compost. While other crops having lesser quantity of farm waste like crop stubbles that can be used to maintain soil organic matter through zero or minimum tillage practices. The addition of organic matter helps in improvement of physico-chemical properties of soil as well as nutrient availability.

Use of Bio-fertilizers

The suitable bio-fertilizer like *Rhizobium* for leguminous crops, *Azospirillum* for grass family crops, while *Azotobactor* used for all type of crops for N-fixation in soil. Moreover, other bio-fertilizers *i.e.* PSB (Phosphorus solubilising bacteria), KSM (Potassium solubilising microorganisms), VAM (Vesicular Arbuscular Mycorrhiza) helps for enhancement of nutrient availability in the root zone. *Rhizobium* culture generally used as seed treatment with juggry solution. However, other bio-fertilizers used as soil application @ 2.5 kg/ha and also used as seedling dip method, sett treatment, tuber treatment, *etc.*

Green Manuring

For improvement of organic matter and fertility status of soil, green manuring is the best option. The green manuring through growing of leguminous crops like *dhaincha*, sunnhemp, cowpea, clusterbean, *etc.*during summer or *kharif* season and incorporated in to the soil at flowering stage or 60 days after sowing helps to added large amount of organic matter, after decomposition released sufficient nutrients for succeeding crop.

Application of Organic Manures

Organic manures are the good source of soil organic carbon as well as all the essential plant nutrients. Organic manures like FYM, farm compost, vermicompost, oil cakes are used for nutrient source for field crops. Bulky organic manures generally applied 15 days before sowing of crops. Commonly FYM @ 10 to 25 tonne/ha used in field crops as nutrient supplement.

Inter-cropping

Cultivation of legume crops in rows with non-legume crops can get additional benefits by supplying biologically fixed nitrogen. Usually clusterbean, cowpea, black gram, green gram, chickpea, *etc.* are grown as inter-crops as per suitable seasons. Leguminous crops enrich the soil by N-fixation and also helps to build-up soil organic matter.

Crop Rotation

The cultivation of legume and non-legume crops in a sequentially on the same plot of land. Crop rotation practice helps to maintain the soil fertility by addition of biologically fixed nitrogen through legume crops that is used for succeeding non-legume crop.

Nutrient Management in Horticultural Crops

Nutrient management is extremely important for fruit and vegetable crops where many time quality of the produce assumes greater consideration than productivity. Application of too much nitrogen, in general, results in oversized, poorly coloured fruit with poor keeping quality. Too little nitrogen also cause problems of poor fruit set, small fruit, pale foliage and stunted growth. Excessive or deficient of other nutrients also can result in serious problem in fruits. Many fruit crops are heavy removers of nutrients and high yield can only be sustained through the application of optimal doses of nutrients in balanced proportion. Diagnosing nutrient deficiency symptoms in fruit and vegetable crops can help in proper nutrient management. Use of soil and or tissue analysis may help to confirm whether symptoms are nutritional or due to some other factors.

Application of Manures

Manures, such as FYM and compost, should be applied or ploughed into the soil 15 days before sowing and transplanting due to the slow release of nutrients from manures. The crop growth is affected if undecomposed or fresh manure is used for cultivation. Therefore, it is advisable to use fully decomposed manure. Generally 15 to 20 tonnes FYM/ha is recommended for the cultivation of vegetable crops, while under organic cultivation, the dose of organic manures calculated on the basis of recommended dose of nitrogen and nutrient content present in organic source.

Application of Bio-fertilizers

Increased fertilizer cost and the awareness of environmental pollution due to fertilizer runoff necessitated the use of bio-fertilizers for the development of more efficient fertility management programme. However, very little work has been done on the use of bio-fertilizers in fruits. The dose and type of bio-fertilizer depends on kind of vegetable crop and the age of fruit plants. Generally bio-fertilizers applied in to soil with mixing in organic manure, while it may be applied through drip irrigation. In case of pomegranate, nitrogen use efficiency has been increased through the use of 25 g *Azotobactor* culture per plant at the time of planting. Similarly, VAM fungi are of the great significance for fruits. It forms a symbiotic association with the roots of plants and helps them in transfer of the nutrients from the soil to the root system. In India, studies on the VAM fungi have so far remained confined to some inoculation trails in some fruits, like citrus, banana, litchi, mango and papaya indicating an increased efficiency of the water and nutrient uptake by host plant. However, more studies are required to adjudge the role of bio-fertilizer in production of the fruits.

Management of the Farm Waste

Biomass is available in the form of leaf fall or plant residue and it may become an integral part of the soil fertility management. The organic farmers, who combine earthworm culture in plant basin with the degradation of such biomass and have come out with some excellent results in sapota, mango, completely eliminating inorganic fertilizer application, are intelligently managing this.

Inter-cropping

To get additional income to the grower until the fruits tree begin to bear and to improve the health of orchard plants, the intercrops are grown which are of the right type. The recommended intercrops in tropical fruit orchards for summer and *kharif* season are clusterbean, cowpea, black gram, green gram, *etc.* While for winter season, the inter crops can grow like pea, chickpea, *etc.* Inter-cropping of pre-bearing orchards with leguminous crops enrich the soil by N-fixation and also build-up organic matter by incorporation of leftover crop residues.

Green Manuring

The cultivation of green manure crops like *dhaincha*, sunnhemp, *etc.* during pre-bearing of fruit orchards and incorporated in to soil as *in-situ* is a source of organic nutrient supplement.

Mulching

Generally mulching is used for prevention of soil moisture loss, weeds and to maintain soil temperature. While organic mulching material used in fruit orchards also helps to addition of organic matter in the soil that is supplement source of nutrients.

Fertigation

The application of water soluble fertilizers applied through irrigation is called fertigation. Generally under drip irrigation system in horticultural crops, the water soluble organic fertilizers like liquid bio-fertilizers, organic nutrient formulations, cow urine, *jivamrut*, waste decomposer, *etc.* can be applied with irrigation water to correct the multi-nutrient deficiency in standing crops.

Foliar Spray of Organics

The applications of nutrient by dissolving in lower concentration and apply on the foliage of crop plants to correct the nutrient deficiency and also work as bio-enhancer. Generally some organics applied through foliar spray like cow urine @5 per cent, *panchgavya* @3 per cent and vermiwash @ 10 per cent and it can be repeats whenever required.

REFERENCES

Hansen, A.L. (2010). *Organic Farming Manual*: A Comprehensive Guide to Starting and Running a Certified Organic Farm, Storey Publishing LLC.

Jnyanendra Kumar Prusty, Sanjaya Kumar Patro and S.S. Basarkar (2016). Concrete using agro-waste as fine aggregate for sustainable built environment – A review, *International Journal of Sustainable Built Environment.*

Palaniappan, S.P. and Annadurai, K. (2018). *Organic Farming*: Theory and Practice, Scientific Publisher.

Panda, H. (2013). *Integrated Organic Farming Handbook*, Asia Pacific Business Press Inc.

Panda, S.C. (2013). *Principles and Practices of Organic Farming*, Agrobios.

Singh, H.P. and Thomas, G.V. (2010). *Organic Horticulture*: Principles, Practices and Technologies, Westville Publishing House.

Yadav, J.S.P. and Singh, G.G. (2000). Natural resource management for agricultural productivity in India, *Indian society of Soil Science*, New Delhi, India.

Chapter 7

Nutrient Management in Organic Cultivation of Coriander (*Coriandrum sativum* L.)

Bindesh Arya, K.C. Meena*, D.K. Patidar, Nitin Soni and B.K. Patidar

College of Horticulture, Rajmata Vijayaraje Scindia Krishi Vishwa Vidyalaya, Mandsaur–458001, Madhya Pradesh
**e-mail: drkailashmeena06@gmail.com*

ABSTRACT

The experiment was laid out in the department of Plantation, Spices, Medicinal and Aromatic Crops, College of Horticulture, RVSKVV, Mandsaur (M.P.) during rabi season in the year of 2019-2020. The experiment was designed in factorial randomized block design with three replications. Coriander (Coriandrum sativum L.) is one of the most important seed spices crop. Vermicompost is a finely divided peat-like material with high porosity, aeration, drainage, and water-holding capacity and usually it contains most plant nutrients in the available form. Bio-fertilizers with organic manures such as Azotobacter, Azospirillum and PSB has decreased the use of chemical fertilizers and have available high quality organic product which is free from harmful agrochemicals for the safety of public health. The present study was carried out in two variety of coriander with the soil application of different organic manure and biofertilizers combinations along with one untreated control plot as recommended dose of fertilizers. Coriander variety Arc 1 performed batter as compared to Acr 2 with the application of vermicompost and biofertilizers among all the morphological, growth and quality parameters.

Keywords: *Azotobacter, Azospirillum, PSB, Organic manures and Coriander.*

Overview of Coriander and Organic Manures

Coriander (*Coriandrum sativum* L.) is one of the important seed spices crop.

Its center of origin is Mediterranean region and belongs to the family Apiaceae. The plant is a smooth, erect annual herb, 30-90 cm high, with conspicuously enlarged nodes and hollow internodes. The fruit is a schizocarp, globular, yellow in color with brown ribs. All the above ground and also of coriander seeds are used in a number of culinary preparations (Jhankar. *et al.*, 2017)and as an aromatic as well as medicinal herb. The major coriander growing states includes Rajasthan, Madhya Pradesh, Andhra Pradesh, Gujarat, Bihar, Uttar Pradesh and some pockets in Tamil Nadu, Odisha, Karnataka and Haryana. In Madhya Pradesh the major coriander growing districts are Guna, Shivpuri, Mandsaur, Neemuch, Ratlam, Shajapur, Rajgarh and Vidisha.

Vermicompost is a product of degradation of the organic matter through the interactions between earthworms and other microorganisms. It is a finely divided peat-like material with high porosity, aeration, drainage, and water-holding capacity and usually it contains most plant nutrients in the available form such as nitrates, phosphates, exchangeable calcium and soluble potassium (Darzi *et al.*, 2015). Bio-fertilizers with organic manures such as *Azotobacter*, *Azospirillum* and PSB, has decreased the use of chemical fertilizers and have available high quality organic product which is free from harmful agrochemicals for the safety of public health. The free-living nitrogen fixing bacteria were found to show not only the ability to fix the atmospheric nitrogen but also the capability to release plant hormones which are similar to gibberellic acid and indole acetic acid in action and could motivate plant growth, absorption of nutrients, and photosynthesis (Mahfouz *et al.*, 2007). A balance fertilizer dosage with FYM, vermicompost and bio-fertilizer as per the crop type is required to be standardized which helps to compute the entire nutrient requirement of the crop throughout its growth period which culminates into higher yields (Jhankar *et al.*, 2017).

Substantial and Approaches During Research

The experiment was laid out at the department of Plantation, Spices, Medicinal and Aromatic Crops, College of Horticulture, Mandsaur under Rajmata Vijayaraje Scindia Krishi Vishwa Vidyalaya, Gwalior (M.P.) during *rabi* season in the year of 2019-2020. The present experiment was designed in factorial randomized block design with three replications. Mandsaur is situated in Malwa plateau in western part of Madhya Pradesh at north latitude of 23.45° to 24.13° and 74.44° to 75.18° east longitude and at an altitude of 435.02 meters above mean sea level. This region falls under agro-climatic zones number10 of the state. The present study was carried out in two variety of coriander such as V_1-Acr 1 and V_2-Acr 2 with the soil application of different organic manure (vermicompost @3 T ha^{-1}+ FYM@7 T ha^{-1}) and biofertilizers (*Azotobacter* + *Azospirillum* + PSB@ 2.5 kg ha^{-1}) combinations along with one untreated control plot as recommended dose of fertilizers. All the morphological and growth parameters were recorded at 30, 60, 90 days after sowing and at harvest and were statistically analyzed and its variance for these data has been furnished results and discussion. Observations were recorded under investigation *i.e.* plant height, number of branch per plant, number of umbels, number of umbellets, seed yield, biological yield, harvest index and essential oil content in seed and dry matter content. Interpretation of data has been made on the mean basis Statistical analysis. The present experiment was carried out using three replicates each in factorial randomized block design. Results were expressed as mean ± standard error (SE) and the data were analysed statistically by two-way ANOVA using SPSS software (version 16.0).

Outcomes and Discussion

All the morphological parameter, yield parameters and quality parameters of both the varieties were significantly influenced with the application of organic manures and biofertilizers. The result confirm that, among the varieties, V_1 Acr 1 (8.66, 46.77, 104.88 and 106.81 cm) had the maximum plant height as compared to V_2 Arc 2 (7.93, 44.74, 101.82 and 104.93) which was lowest in the same trait. Whereas, in organic manures and biofertilizers, OM_1 vermicompost @3t + biofertilizers @ 2.50 kg ha^{-1} (8.94, 47.11, 104.96 and 107.59 cm) attended more plant height and OM_0 RDF(7.55, 43.41, 100.91 and 102.72) less in plant height. Moreover, in the interactions V_1 x OM_1 (9.57, 48.04, 107.29 and 108.98) also higher and V_2 x OM_0 (7.66, 43.00, 100.28 and 102.68) lower in the same trait at 30, 60, 90 days after sowing and at harvest respectively (Figure 7.1). The better escalation in plant height with spread may be due to the build-up of colonies of the applied bio-fertilizers inoculates and organic manures as mentioned. This may be because of better phosphorus solubilization due to accelerated bacterial activity (Hnamte *et al.*, 2013) in anise and also improving the soil environment which in turn encourages proliferous root growth which might have resulted in more nitrogen accumulation, absorption of moisture, nutrients and rate of photosynthesis thus, producing higher biomass (Singh, 2011), which also is in agreement with Jhankar *et al.* (2017) in coriander. Similar, findings were also confirmed with the outcomes of Sajeena *et al.* (2018) in *Pimpinella anisum*. From the Figure 7.2 the maximum accumulation of number of branches was noted in the variety V_1Acr 1(4.59, 7.43 and 8.18) though minimum in V_2 Acr 2(3.94, 6.58 and 7.44). While, in organic manures and biofertilizers, OM_1 vermicompost @3t + biofertilizers @ 2.50 kg ha^{-1} (4.72, 7.46 and 8.35) had accrued higher and OM_0RDF (3.61, 6.24 and 6.81) lower in the same. Furthermore, in the interactions V_1x OM_1 (5.24, 7.91 and 8.7) also higher and V_2x OM_0 (3.54, 6.05 and 6.64) lower in the same trait at 60, 90 days after sowing and at harvest respectively. Organic manure applications encouraged the metabolic process of the plant, which triggered in faster cell elongation, division and multiplication in meristematic region of the plant due to production of plant growth substances and other growth promoting substances and this is the reason of higher number of branches. This finding are similar with the result of Hnamte *et al.* (2013), Kumar *et al.* (2015) and Jhankar *et al.* (2017) in coriander. It is confirmed from Figure

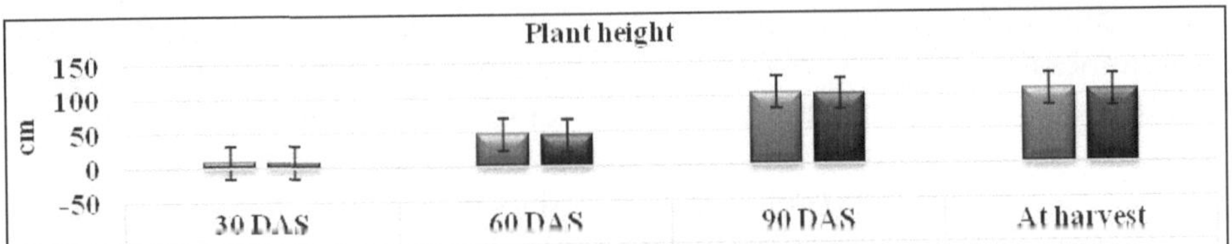

	30 DAS	60 DAS	90 DAS	At harvest
Acr 1	8.67	46.77	104.64	106.81
Acr 2	7.94	44.74	102.06	104.93

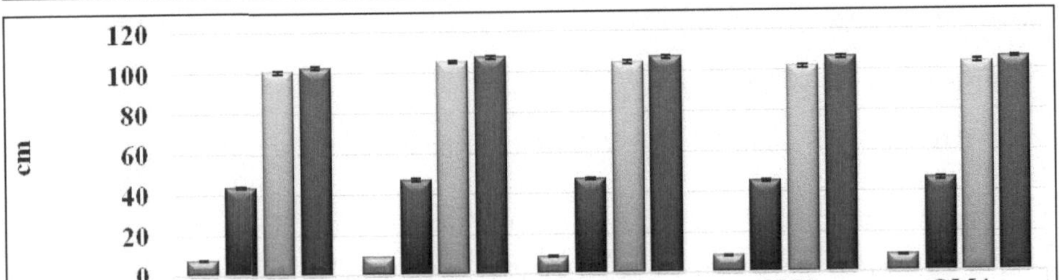

	OM0	OM1	OM2	OM3	OM4
30 DAS	7.55	8.95	8.55	8.14	8.33
60 DAS	43.41	47.12	46.86	45.27	46.13
90 DAS	100.91	105.39	104.48	102.02	103.97
At harvest	102.73	107.59	106.85	106.25	105.95

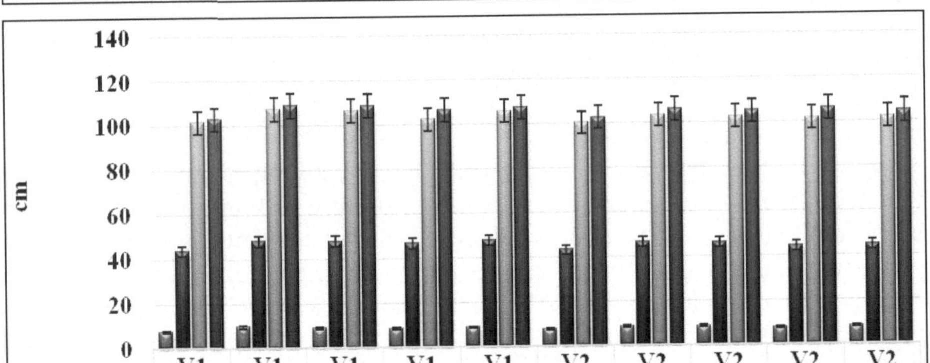

	V1 OM0	V1 OM1	V1 OM2	V1 OM3	V1 OM4	V2 OM0	V2 OM1	V2 OM2	V2 OM3	V2 OM4
30 DAS	7.66	9.57	8.84	8.61	8.67	7.45	8.32	8.27	7.68	7.99
60 DAS	43.82	48.04	47.74	46.55	47.70	43.00	46.28	45.98	43.99	44.56
90 DAS	101.53	107.29	106.32	102.29	105.78	100.29	103.49	102.63	101.74	102.17
At harvest	102.78	108.99	108.73	106.39	107.18	102.68	106.20	104.97	106.12	104.71

Figure 7.1: Responses of Varieties, Organic Manures and Biofertilizers on Plant Height.

7.3 that, among the varieties, V_1Acr 1(28.77) noted the highest number of umbels and V_2Acr 2(25.67) lowest in the same trait. However, in organic manures and biofertilizers OM_1 vermicompost @3t + biofertilizers @ 2.50 kg ha^{-1}(29.80) attended maximum and OM_0 RDF(22.36) minimum in number of umbels.

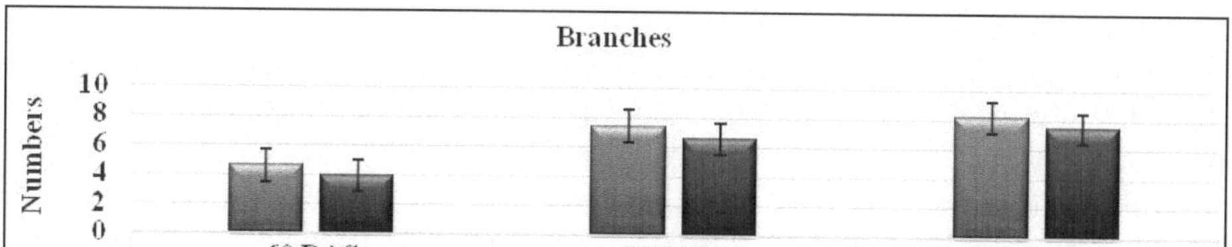

	60 DAS	90 DAS	At harvest
Acr 1	4.6	7.43	8.18
Acr 2	3.95	6.58	7.44

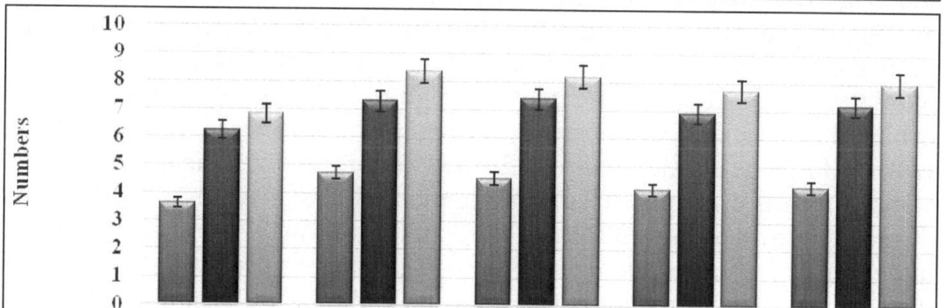

	OM0	OM1	OM2	OM3	OM4
60 DAS	3.61	4.73	4.55	4.18	4.29
90 DAS	6.24	7.29	7.4	6.91	7.2
At harvest	6.82	8.35	8.19	7.72	7.98

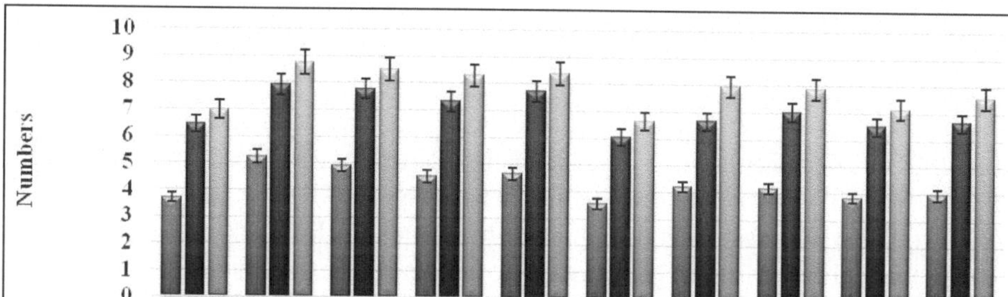

	V1 OM0	V1 OM1	V1 OM2	V1 OM3	V1 OM4	V2 OM0	V2 OM1	V2 OM2	V2 OM3	V2 OM4
60 DAS	3.68	5.24	4.92	4.53	4.63	3.55	4.21	4.17	3.84	3.96
90 DAS	6.43	7.91	7.78	7.32	7.73	6.05	6.66	7.02	6.5	6.66
At harvest	6.99	8.75	8.49	8.29	8.39	6.64	7.95	7.89	7.15	7.57

Figure 7.2: Responses of Varieties, Organic Manures and Biofertilizers on Branches.

Moreover, under the interactions, V_1x OM_1 (31.33) recorded the maximum andV_2x OM_0 (21.52) minimum numbers. Application of vermicompost with biofertilizers has improvement of biological activities of soil and mineral element absorption which, leads to more biomass production and subsequently enhanced number of umbels per plant. Darzi *et al.* (2015) has suggested that, vermicompost increases the growth rate and yield attributing parameters through gradual mineralization, water and mineral uptake in coriander. Our findings are in agreement with the comments of in coriander. The higher number of umbellate found in variety V_1 Acr 1(6.83) as compared to V_2Acr 2 (6.52) which was lowest in the same trait. Whereas, in organic manures and biofertilizers, OM_1 vermicompost @3t + biofertilizers @ 2.50 kg ha^{-1}(6.84) attended maximum and OM_0 (6.43) minimum in number of umbellate per plant. Moreover, under the interactions, V_1 x OM_1 (7.06) also higher and V_2 x OM_0 (6.40) lower in the same trait (Figure 7.3). The result might be due to the balanced nutrition of biofertilizers and organic manures in establishment of proliferous root system and higher absorption of nutrients and water with enhanced physical environment. Numbers of umbellate per umbel were highest with the combine application of biofertilizers and organic manures. The present results are in close accordance with the findings of Darzi *et al.* (2015) and Patidar *et al.* (2016) in coriander. The result confirms that (Figure 7.6), among the varieties, V_1Acr 1(11.05) attended the maximum seed yield and V_2Acr 2(9.67) minimum in the same trait. Whereas, under organic manures and biofertilizers OM_1 vermicompost @3t + biofertilizers @ 2.50 kg ha^{-1}(11.48) attended maximum and OM_0RDF (8.79) minimum in seed yield. Moreover, in the interactions V_1 x OM_1 (12.03) also higher and V_2 x OM_0 (8.73) lower in the same trait. Increased seed yield in treated plots may be owing to the improvement of yield attributing components *viz.*, number of umbels, number of umbellate, number of seed per umbel and other components as well as total biomass yield. Organic manures and bio-fertilizers increased the supply of mineral elements, through vermicompost application, which leads to higher absorption and utilization of these elements having direct effect on the yield attributes as well as the grain yield according to Vadiraj *et al.* (1998) and Darzi *et al.* (2015) in coriander. Our finding is in accordance with comment of Valiri *et al.* (2015).The results had indicated that (Figure 7.3), among the varieties, V_1Acr1(19.48) was assessed the maximum harvest index as compared to V_2Acr 2(20.37) which was lowest. Whereas, now organic manures and biofertilizers OM_1 vermicompost @3t + biofertilizers @ 2.50 kg ha^{-1}(19.51) attended maximum and OM_0RDF (20.21) minimum harvest index. Moreover, in the interactions V_1 x OM_1 (18.18) higher and V_2 x OM_0 (20.57) lower in the same trait. The better growth in areal part of the plant with their spread may be due to the applications of organic manures which, encouraged the production of plant growth substances and other growth promoting substances. The biofertilizers improved phosphorus solubilization due to accelerated bacterial activity. Resulted in more nitrogen accumulation, absorption of moisture, nutrient and accelerate rate of photosynthesis, which leads to the higher dry matter production and harvest index according to Singh *et al.* (2011), Hnamte *et al.* (2013) and (Shirkhodaei *et al.*, 2014). This findings are similar with the result of (Kumar *et al.*, 2015) and (Jhankar *et al.*, 2017) and in coriander. Among the varieties, V_1Acr 1(56.93) had the maximum biological yield as compared to V_2Acr 2(47.49) which was lowest in the same trait. Whereas, in organic manures and biofertilizers OM_1

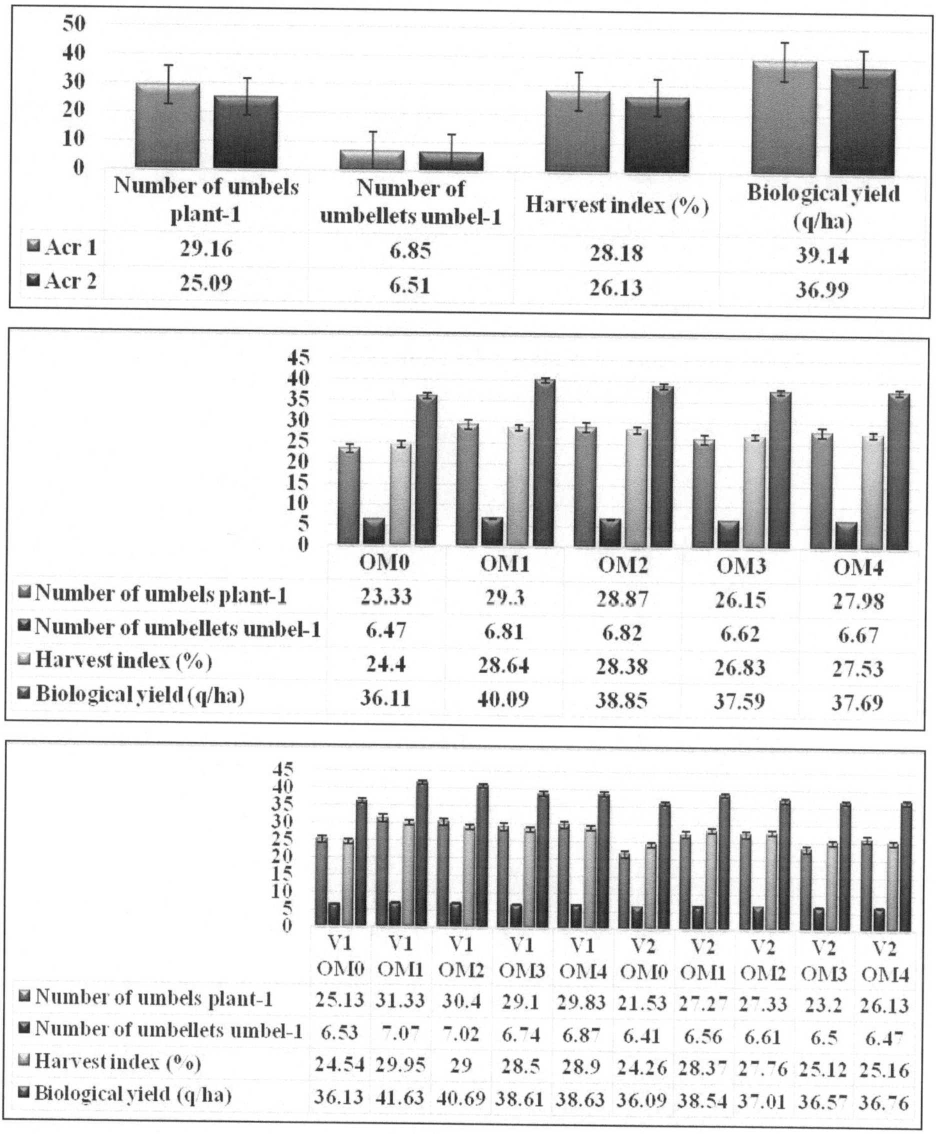

	Number of umbels plant-1	Number of umbellets umbel-1	Harvest index (%)	Biological yield (q/ha)
Acr 1	29.16	6.85	28.18	39.14
Acr 2	25.09	6.51	26.13	36.99

	OM0	OM1	OM2	OM3	OM4
Number of umbels plant-1	23.33	29.3	28.87	26.15	27.98
Number of umbellets umbel-1	6.47	6.81	6.82	6.62	6.67
Harvest index (%)	24.4	28.64	28.38	26.83	27.53
Biological yield (q/ha)	36.11	40.09	38.85	37.59	37.69

	V1 OM0	V1 OM1	V1 OM2	V1 OM3	V1 OM4	V2 OM0	V2 OM1	V2 OM2	V2 OM3	V2 OM4
Number of umbels plant-1	25.13	31.33	30.4	29.1	29.83	21.53	27.27	27.33	23.2	26.13
Number of umbellets umbel-1	6.53	7.07	7.02	6.74	6.87	6.41	6.56	6.61	6.5	6.47
Harvest index (%)	24.54	29.95	29	28.5	28.9	24.26	28.37	27.76	25.12	25.16
Biological yield (q/ha)	36.13	41.63	40.69	38.61	38.63	36.09	38.54	37.01	36.57	36.76

Figure 7.3: Responses of Varieties, Organic Manures and Biofertilizers on Umbels, Umbellate, Harvest Index and Biological Yield.

vermicompost @3t + biofertilizers @ 2.50 kg ha^{-1}(59.33) attended maximum and OM_0 RDF (43.59) minimum. Moreover, in the interactions $V_1 \times OM_1$ (66.18) also higher and $V_2 \times OM_0$ (42.44) lower in the same trait (Figure 7.3). The results clearly demonstrated the effectiveness of vermicompost in increasing the biological yield. Vermicompost increases the growth rate because of water and mineral uptake such as; nitrogen and phosphorus, which lead to the biological yield improvement. Effect of phosphate solubilizing bacteria on the biological

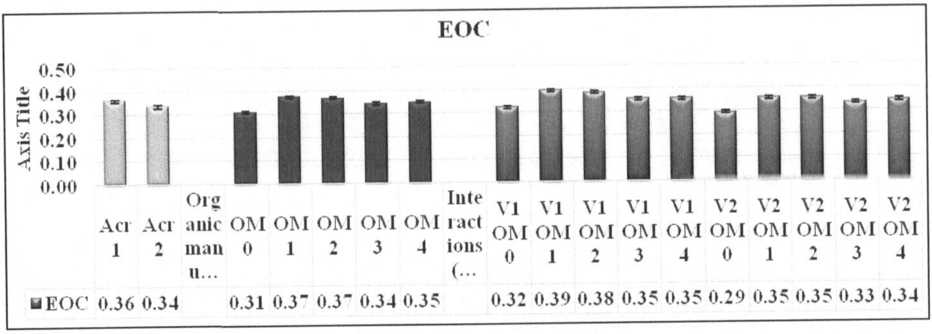

Figure 7.4: Responses of Varieties, Organic Manures and Biofertilizers on Plant Height Essential Oil Content in Seed.

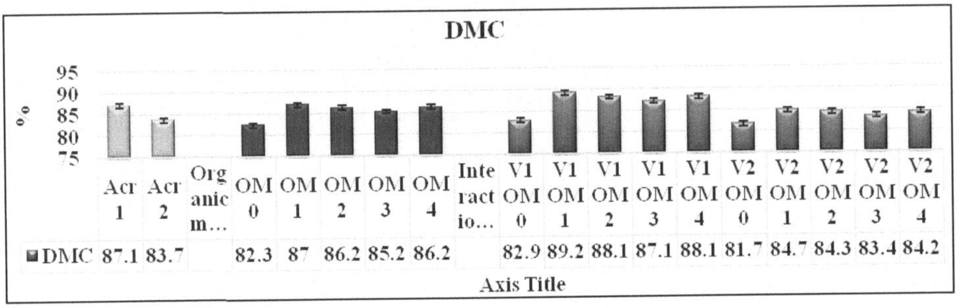

Figure 7.5: Responses of Varieties, Organic Manures and Biofertilizers on Dry Matter Content in Seed.

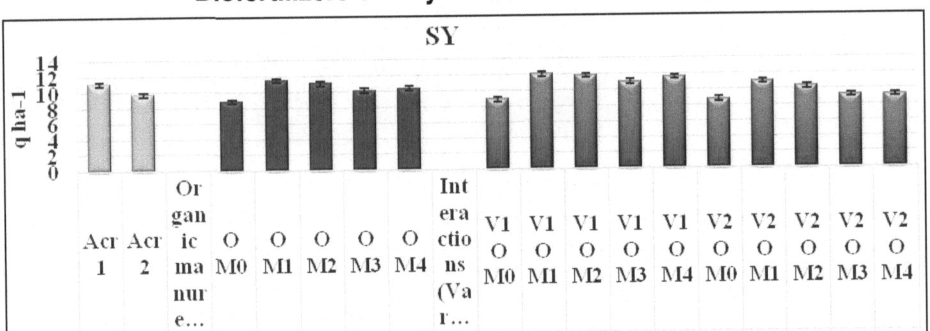

Figure 7.6: Responses of Varieties, Organic Manures and Biofertilizers on Seed Yields.

yield was due to increased phosphorus uptake. All the earlier reports and present results supported the fact of positive and synergistic effect on interaction between two factors which highly dependent on the effect of organic matter, containing vermicompost, on the activity of phosphate solubilizing bacteria. The result of the present work is in agreement with the reports of Darzi *et al.* (2012) on anise. Essential oil content in seed was significantly influenced by the application of organic manures and biofertilizers. The findings exhibited that

(Figure 7.4), among the varieties; V_1Acr 1(0.36) had the maximum essential oil content in seed as compared to V_2Acr 2(0.34) which was lowest in the same trait. Whereas, in organic manures and biofertilizers OM_1vermicompost @3t + biofertilizers @ 2.50 kg ha^{-1}(0.37) attended maximum and OM_0 RDF (0.31) minimum in essential oil content in seed. Moreover, in the interactions V_1 x OM_1 (0.39) also higher and V_2 x OM_0 (0.29) lower in the same trait. Thought, application of vermicompost with biofertilizer were increased of the mineral uptake such as growth substances and other essential elements has a positive effect on proper biomass production and subsequently the enhanced essential oil content according to Darzi *et al.* (2013) in coriander. The result confirm that (Figure 7.5), among the varieties, V_1Acr 1(87.08) had the maximum dry matter content in seed as compared to V_2 Acr 2(83.65) which was lowest in the same trait. Whereas, in organic manures and biofertilizers OM_1vermicompost @3t +biofertilizers @ 2.50 kg ha^{-1} (86.95) attended maximum and OM_0 RDF (82.28) minimum in dry matter content in seed. Moreover, in the interactions V_1 x OM_1 (89.21) also higher and V_2 x OM_0 (81.67) lower in the same trait. Dry matter production also showed significant variation among the different biofertilizer and micronutrient treatments. Among the biofertilizer and organic manure recorded maximum dry matter production. The application of biofertilizers might have enhanced the availability of nutrients and the production of growth hormones could have contributed to the increase in the dry matter production. Similar results were obtained by Mehta *et al.* (2011), Mounika *et al.* (2017) in coriander, Kusuma *et al.* (2019).

CONCLUSION

On the basis of one year research it could be concluded that, out of 10 treatment combinations, the vermicompost @ 3 T ha^{-1}+ biofertilizers (Azotobacter and PSB @ 2.5 kg ha^{-1}) are best performing combination of organic manures and biofertilizers for maximizing seed yield and enhancement of quality in coriander.

REFERENCES

Darzi, M.T., Seyedhadi, M.H. and Rejali, F. 2012. Effects of the application of vermicompost and phosphate solubilizing bacterium on the morphological traits and seed yield of anise (*Pimpinella anisum L.*). *J. Medicinal Plants Res.*, 6(2): 215-219.

Darzi, M.T., Shirkhodaei, M. and Hadi, M.H.S. 2013. Effects of Vermicompost and Azotobacter and Azospirillum bacteria on quantity and quality of essential oil of coriander (*Coriandrum sativum* L.). *Int.J. of Farming and Allied Sci.* 2: 1277-1283.

Darzi, M.T., Shirkhodaei, M. and Hadi, M.R.H.S. 2015. Effects of vermicompost and nitrogen fixing bacteria on seed yield, yield components of seed and essential Oil content of Coriander (*Coriandrum sativam*). *J. Medicinal Plant and By–products*, 1: 103-109.

Hnamte, V., Chatterjee, R. and Tania, C. 2013. Growth, flowering, fruit setting and maturity behavior of coriander (*Coriandrum Sativum* L.) with organics including biofertilizers and inorganic. *The Bioscan* 8(3): 791-793.

Jhankar, P., Panda, C.M. and Sethi, D. 2017. Effect of INM practices on yield, yield attributes and economics of Coriander (*Coriandrum sativam* L.). *Int. J. Curr. Microbiol. App. Sci.* 6(5): 1306-1312.

Kumar, R., Singh, M.K., Kumar, V., Verma, R.K., Kushwah, J.K. and Pal, M. 2015. Effect of nutrient supplementation through organic sources on growth, yield and quality of coriander (*Coriandrum sativum* L.). *Indian J. Agric. Res.* 49: (3) 278-281.

Kusuma, M.V., Venkatesha, J., Ganghadarappa, P.M., Hiremath, J.S., Mastiholi, A.B. and Manjunatha, G. 2019. Effect of Integrated Nutrient Management on Growth and Yield of Fennel (*Foeniculum vulgare* Mill.). *Int. J. Curr. Microbiol. App. Sci.* 8(1): 2782-2794.

Mahfouz, S.A. and Sharaf Eldin, M.A. 2007. Effect of mineral vs. biofertilizers on growth, yield, and essential oil content of fennel (*Foeniculum vulgare* Mill). *Int. J. Agrophis*. 21: 361-366.

Mehta, R.S., Patel, B.S. and Jat, R.A. 2011. Effect of Nitrogen, Phosphorus and Bio-Fertilizer inoculation on Growth, Productivity, Nutrient Uptake and Economic Returns in Fenugreek (*Trigonella foenum-Graecum* L). *Ind. J. Dryland Agric. Res. Dev.* 26(1): 102-108.

Mounika, Y., Sivaram, G.T., Reddy, P.S.S and Ramaiah, M. 2017. Effect of biofertilizers and micronutrients on growth, leaf yield and quality of coriander (*Coriandrum sativum* L.) cv. Sadhana. *J. Hort. Sci.* 12(2): 113-117.

Singh, M. 2011. Effect of vermicompost and chemical fertilizers on growth, yield and quality of coriander (*Coriandrum sativam* L.) in a semi-arid tropical climate. *J. Spices and Aromatic Crops*, 20(1): 30-33.

Shirkhodaei, M., Darzi, M.T.and Hadi, M.H.S. 2017. Influence of Vermicompost and Biostimulant on the growth and biomass of coriander (*Coriandrum sativum* L.). *Int. J. Advanced Biological and Biomedical Res.* 2: 706-714.

Sanjeeva, M.B., Rao, G.G. E. and Venkatachalapathi. V. 2018. Effect of Sources and Levels of Organic Manures on Growth and Yield of Black Cumin. *Int. J. of Science and Res.*, 2319-7064.

Patidar, L., Ranjan, J.K., Singa, B., Mishra, B.K., Aiswath, O.P., Kant, K., Sharma, B. and Rai, R.K. 2016. Influence of integrated supply of AM, PSB, Azotobacter and inorganic fertilizer on growth, yield and quality in coriander (*Coriandrum sativum*) and micro-flora population in the soil. *Ind. J. of Agril. Sci.* 86(9): 1140-4.

Vadiraj, B.A. and, Potty, S.N. 1998. Response of coriander (*Coriandrum sativum* L.) cultivars to graded levels of vermicompost. *J. Spices and Aromatic Crops.* 7(2): 141-143.

Valiki, S.R., Ghanbari, H.S., Golmohammadzadeh, S. and Tat, O.F. 2015. The Effect of Vermicompost and NPK Fertilizer on Yield, Growth Parameters and Essential Oil of Fennel (*Foeniculum vulgare*). *Int. J. Life Sci.* 9(4): 38– 43.

Chapter 8

Diseases and Insect Pests Management of Vegetable Crops through Biological Agents

R. Chandra[1], Sachin Kumar Jain[2]* and D. Ram[3]

*[1]Department of Horticulture, [2]Department of Plant Pathology,
Amar Singh College, Lakhaoti, Bulanshshahr–203407, Uttar Pradesh
[3]Department of Horticulture, Acharya Narendra Deva University of Agriculture and Technology Kumarganj, Ayodhya–224229, Uttar Pradesh
e-mail: sachinjain1115@gmail.com

ABSTRACT

Insect pests are the most destructive constraints to world agriculture. They damage millions of hectares of crops belonging to food grains, horticulture crops every year. The extent of damage is increasing year after year under the influence of changing climatic conditions. The developing economies suffer more due to outbreak of insect pests in fruits and vegetables which are the major source of international trade. Insect pests create a major barrier to production and productivity as they are causing several outbreaks. Chemical pesticides are being used to manage these insect pests in order to achieve quick control. But, in many cases, misuse of pesticides led to environmental pollution along with population resurgence, pesticide residue problem in soil and water, and pest resistance to these chemical pesticides. In such situation, biological control gives a promising hope towards sustainable pest management. Biological control is the usage of living organisms and their products alternative to chemical pesticides or integrating as one of the component in integrated pest management program. Here, we can employ various biological control agents like natural enemies (predators and parasitoids), entomopathogenic fungi, bacteria, virus, nematodes, protozoa and actinomycetes or their products for effective pest management. They can be directly used or formulated into a commercial formulation suitable for various methods of application. Classical biological control is well-tried, cost effective approach to manage various pests in vegetable crops. Since we use native strains of bio-control agents, they will be cost effective, more effective due to acclimatization to introduced microclimate and keeps the pest level below acceptable economic threshold levels. In order to reduce the serious problems of chemical pesticides on environment and human life bio-control is suitable approach. Vegetables being high value, low volume crops and shares major portion of our daily diet needs to be grown pest free and residue free. In order to achieve sustainable pest control and reduce cost of protection, biological control is well suited for management of insect pests in vegetable crops.

Keywords: *Diseases, Insect pest, Bio-control, Vegetables.*

INTRODUCTION

In India, the estimated gross loss in crop production due to insects and diseases is staggering 15 per cent or 290 billion per annum. On the other hand, over use of chemical pesticides on crops causes deleterious effects through accumulations of toxic residue in food chains, spoiling soil fertility and natural biodiversity, and finally resulting in adverse effects on the plants, soil and human health. In view of the above, an alternative to toxic chemical pesticides, bio-management (an eco-friendly, cost effective and easy to handle) of insect pests and diseases evolved and is being used for sustainable agriculture (Annadurai and Palaniappan, 2018, Prakash, 1997, Haldhar *et al.*, 2017, Singh, 2012). This chapter is aimed to provide meaningful information about bio-management of diseases and insect pests of vegetable crops under following head:

1. Mass production of fungal bio-control agents
2. Preparation of oilseed cake
3. Preparation of botanical antagonists and
4. Applications of bio-pesticides
5. Mode of Action of Bio-pesticide
6. Advantages of Bio-management

Mass Production of Fungal Bio-control Agents

Management of fungal bio-control agents like *Trichoderma viride, T. virens, T. harzianum, Aspergillus niger, A. terreus, Cladosporium oxysporum, Paecilomyces lilacinus, Beauveria bassiana* and *Metarhizium anisopliae, etc.* (known for their fungicidal, insecticidal, nematicidal activities) for the mass production of bio-control agents are most important step in the bio-management of diseases and pests of vegetable crops. Above bio-control agents can be isolated from the locality or procure from public institutions (like IARI, *etc.*) involved in isolation and characterization of bio-control agents. For mass production of collected bio-control agents, overnight soaking of starch rich grains of sorghum in water followed by draining out and filling about 100 - 150 gram of the same soaked grains in auto cleavable polypropylene bags (packets). The bags packed with grains are sealed with thread and sterilized in pressure for 35-40 minutes. (About 4-5 whistles). After cooling, around 10-15 pre-mycelium colonized grains from already prepared bio-control agents, transferred in each fresh packet (containing sterilized grains) and packets are sealed quickly. Transfer of pre-mycelium colonized grains in the packets and sealing of packets are done near the flame of the lamp. Inoculated bags were incubated (kept) for 10-12 days at ambient temperature (20-25 °C) for till profuse growth of bio-control agents following which they were ready to apply in the fields (Goswami and Pandey, 2008).

Preparation of Oilseed Cake

Oilseed cake of castor (*Ricinus communis*), neem (*Azadirachta indica*), mahua (*Madhuca Indica*), karanj (*Pongamia glabra*), *etc.* in addition to their fungicidal and nematicidal properties, have also been reported to be ideal medium for the proliferation of fungal bio-control agents and help in the plant growth due to their rich constituents of NPK. They also make plants more tolerant against plant diseases. Oil seed cake can also be easily extracted through speller by crushing the dried seeds in the village levels or purchase from the local market.

Preparation of Botanical Antagonists

The most important botanical antagonists are kalmesh (*Andrographis paniculata*), oak (*Calotropis gigantea*), dhatura (*Datura stramonium*), bhangra (*Eclipta alba*), jungli aonla (*Phyllanthus niruri*), satyanashi (*Argemone mexicana*), undi (*Calophyllum inophyllum*), *etc.* Botanical antagonists are prepared by collection of above plants, drying of collected plants in the shade, powdering of dried plants and packing power in bags.

Applications of Biopesticides

Above prepared bio-pesticides can be used for the effective bio-management of soil borne fungal diseases, root knot nematodes, foliar diseases caused by fungi and insect pests. For the bio-management of diseases and insect pests of vegetable crops, following doses of bio-pesticide can be used:

Dose of Biopesticide for the Treatment of Seeds

For the treatment of one kg vegetable seeds with bio-pesticide, following amounts of bio-constituents are weighted on the electric balance:

Trichoderma species 8-10g,

Paecilomyces lilacinus 8-10g,

Neem/karanj oil seed cake 50g and

Botanical antagonists 50g

Above weighted amounts of bio-constituents are dissolved in 500 ml water along with sticker, *e.g.*, jaggery (50g). Prior to the sowing of seeds in the field, seeds are coated (mixed) with the above prepared solution of bio-pesticide and dried in shade. Now above seeds can be used for sowing in the field.

Dose of Bio-pesticide for the Treatment of Soil

For the treatment of one hectare soil with bio-pesticide, following amounts of bio-constituents are weighted on the electric balance:

Trichoderma species 2 kg

Paecilomyces lilacinus 2kg

Neem/karanj oilseed cack 200kg and

Botanical antagonists 20kg

Above weighted amounts of oil seed cake and botanical antagonists are mixed together and broadcast in the field. Field is irrigated with water for the fast decomposition oil seed cake and botanical antagonists. After 10 days, above fungal bio-control agents along with farm yard manure (FYM) to be applied at the time of sowing of seeds/transplanting in the furrow as soil treatment.

Dose for Root Dip Treatment of Seedlings

For the treatment of roots of seedlings with bio-pesticide, following amounts of bio-constituents are weight on the electric balance:

Trichoderma species 8-10g,

Paecilomyces lilacinus 8-10g,

Neem/karanj oilseed cake 50g and

Botanical antagonists 50g

Above weighted amounts of bio-constituents are dissolved in one liter water along with sticker, *e.g.*, jaggery (50g). Prior to transplanting of seedlings in the field, roots of seedlings are dipped for 30 minutes in the above prepared solution bio-pesticide.

Doses of Bio-pesticide for the Foliar Broadcasting

For one hectare foliar broadcasting of bio-pesticide, following amounts of bio-constituents are weighted on the electric balance:

Trichoderma species 2 kg

Beauveria bassiana 2 kg

Oilseed cake 10 kg and

Farm yard manure 25 kg

Above weighted amounts of bio-constituents are mixed with 25 kg farm yard manure (FYM) and broadcast in one hectare field. For foliar application through dusting, the above weighted amounts of bio-components to be mixed and after 10 days of seed germination/seedlings transplanting, 4 times broadcasting/dusting to be done at 10 days intervals.

Doses of Bio-pesticide for Field/Foliar Spray

For one hectare field/foliar spray of bio-pesticide, following amounts of bio-constituents are weighted on the electric balance:

Trichoderma species 2 kg,

Beauveria bassiana 2 kg and

Oilseed cake 10 kg

Above weighted amounts of bio-constituents are dissolved in 500 liter of fresh water and sprayed in one hectare field. For foliar spray, the above prepared solution bio-pesticide started spray after 10 days of seed germination/seedlings transplanting. Four sprays can be done at 10 days intervals.

Mode of Action of Bio-pesticide

The combined application fungal bio-control agents, *Trichoderma viride* and *Paecilomyces lilacinus*, is targeted for the management of soil borne wilt and root rot causing fungi while the latter being nematophagous would manage root knot nematode. Botanical antagonists, kateli/satyanashi (*Argemone mexicana*), aok (*Calotropis procera*), jungli aonla (*Phyllanthus niruri*) and oilseed cake in addition to their plant growth enhancing properties, are also reported to be fungicidal, nematicidal and insect repellent. Spraying of fungal bio-control agent, *Beauveria bassiana* (an insect parasite) and neem oil seed cake help in the management of pests through parasitizing and also repelling by fungal bio-control and neem oil seed cake, respectively. Management of pests being also vectors of viruses results in minimizing the incidence of viral diseases (Goswami and Pandey, 2008).

Advantages of Bio-management

The interest in bio-management of diseases and pests of vegetable crops is based on the disadvantages associated with chemical pesticides, some of which are as follows: (i) extensive pollution of the environment, (ii) serious health hazard due to the presence of their residues in the food, fiber and fodder, and (iii) increasing cases of insects developing resistance, *e. g.*, *Helicoverpa* (*Heliothis*) has become resistant to most of the insecticides (Baphna,1997, Dixit, 1997).In view of above, bio-management of diseases and insect pests provide following advantages over chemical pesticides:

1. They protect vegetable crops from nursery to till harvest, against diseases, pests and root knot nematodes infecting a wide range of vegetable crops.
2. They protect vegetable crops from disease complexes caused by root knot nematodes and some pathogenic fungi, the former pre-disposing the crops for the fungal attack.

3. As we know that all the components used for preparing bio-pesticide are eco-friendly (natural products) and they are safe for use by farmers with zero risk.
4. They don't have any adverse effect on the environment.
5. They are much cheaper than chemical pesticides.
6. It results in disease free plants much healthier and thus adds to the yield of the crop.
7. They do not affect the quality of crop produce. Therefore, the produce of crops is safe for consumers.
8. Bio-management of pests and diseases in vegetable crops will reduce the dependency of farmers on costly chemical pesticides.
9. Most of the bio-constituents can be managed at the home or procure from the local market.

Future Prospects

Day by day increasing health awareness among the people certainly enhances the demand for healthy food in the coming years. In this direction, organic food will play an important role in ensuring the good health of the people.

REFERENCES

Annadurai, K. and Palaniappan, S. P. (2018). Organic Farming: Theory and Practice. Scientific Publishers, India.

Baphna, P. D. (1997). My successful experiments in agri-horticulture. Proceedings of National Seminar on the Indian Model for Rural Development in the 21st Century. Organized by Mahamana Malaviya Mission, B. H. U. Unit, Varanasi. pp. 51- 53.

Dixit, P. K. (1997). Some aspects of sustainability of agriculture and of organic farming. Proceedings of National Seminar on the Indian Model for Rural Development in the 21st Century. Organized by Mahamana Malaviya Mission, B. H. U. Unit, Varanasi. pp. 59-62.

Goswami, B. K. and Pandey, R. K. (2008). Bio-management of pests and diseases infecting vegetable crops. Amity Centre for Biocontrol and Plant Disease Management, Amity University, Noida (U. P.)

Haldhar, S. M., Jat, G. C., Deshwal, H. L., Gora, J. S. and Singh, D. (2017). Insect pest and disease management in organic farming. In: Towards Organic Agriculture (Ed. Gangwar, B. and Jat, N. K.). Today and Tomorrow's Printers and Publishers, New Delhi. pp. 359-390.

Prakash, B. (1997).Vichar karen karbnic-krishi ke labhkari pahluo par. Proceedings of National Seminar on the Indian Model for Rural Development in the 21st Century. Organized by Mahamana Malaviya Mission, B. H. U. Unit, Varanasi. pp. 21-22.

http: //www.yogickheti.org/

Singh, B. D. (2012) Biotechnology Expanding Horizons. Kalyani Publishers, New Delhi. pp. 562-565.

Chapter 9

Bio-Intensive Pest Management (BIPM) in Major Vegetable Crops

Abhinav Kumar[1]*, *Krishna Deo Verma*[2] *and Ingle Dipak Shyamrao*[3]

[1]*Department of Entomology, Bhola Paswan Shastri Agricultural College, BAU, Purnia, Bihar*

[2]*Department of Entomology, Amar Singh PG College, Lakhaoti, Bulandshahar, Uttar Pradesh*

[3]*Department of Entomology, Agriculture College Garhwa, Birsa Agricultural University, Ranchi*

**e-mail: abhinavkumar999@gmail.com*

ABSTRACT

Vegetables are considered as the major parts of Indian diet. However, there are several factors that confine its productivity, mainly the biotic factors viz., insect pests and diseases. Bio-intensive pest management (BIPM) is fundamentally a component of conventional integrated pest management. Which will help to lessen our dependency on conventional chemical methods which ultimately leads to ecological deterioration? Bio-intensive pest management (BIPM) essentially have bio-pesticides derived from microbials, parasitoids, predators, botanicals and all other conventional non-chemical methods of pest control. At present our Indian farming, which is experiencing a changeover, is slowly but surely adopting the methods and means of pest management to attain sustainability in agriculture.

Keywords: *Vegetables, Yield losses, Bio-intensive-IPM, Pest management.*

INTRODUCTION

Indian are extensive cultivators and consumers of vegetables; we stand the second largest producer of vegetables after China, about 75 million tons. The existing area under vegetable cultivation in India is nearly around 4.5 million ha (Table 9.1). Majority of Indians are vegetarian, with a per capita consumption 135 g per day as against the recommended 300 g per day. Indian vegetable export is very low because of increased domestic requirement and other limitations in crop production. The key limiting factor, include the wide crop destructions due to augmented pest menace. In various cases, there is 100 per cent yield loss due to viral diseases vectored by insects (Shivalingswami *et al.*, 2002). The extent of crop losses in vegetables varies with the cropping pattern, location, damage potential of the pest involved and cropping season. Vegetables are more prone to various biotic factors mainly due to their tenderness and softness as compared to other crops and most importantly virtual absence of resistance characters because of intensive hybrid cultivation. The insect pests cause crop losses up to 40 per cent in vegetable production (Srinivasan, 1993) (Table 9.2). We make known to high yielding varieties and hybrids, in order to bridge the ends of increased demand and limited productivity. This has caused in intense changes in pest situation leading to minor pests assuming major status, like serpentine leaf miner *Liriomyza trifolii* (Burgess), which has become a serious problem in tomato hybrid growing pockets. Vegetable growers by and large depend on chemical pesticides to counter the problem of insect pests. It accounts for 13-14 per cent of total pesticides consumption, as against 2.6 per cent of cropped area (Sardana 2001). Indiscriminate use of pesticides has led to severe ecological consequences like destruction of natural enemy fauna, effect on non-targeted organisms, residues in consumable products including packed pure and mineral water and eventually resistance to the pesticides, to which we solely rely. The Bio-intensive pest management (BIPM) concept is the recent trend in Indian farming and attracting the farmers for higher income to their produce. This has

Table 9.1: Area under Vegetable Cultivation in India

Sl.No.	*State*	*Production (in '000 MT)*
1	Uttar Pradesh	28,316.45
2	West Bengal	27,695.29
3	Madhya Pradesh	17,545.48
4	Bihar	15,863.21
5	Maharashtra	12,306.72
6	Gujarat	12,254.29
7	Odisha	8,766.82
8	Karnataka	8,394.15
9	Haryana	7,151.66
10	Chhattisgarh	7,003.25

Table 9.2: Yield Losses Due to Insect Pests in Vegetables

Crop Pests	*Yield Loss (Per cent)*	*States*
Cabbage		
Diamondback moth (*Plutella xylostella*)	16.87-98.83	Karnataka
Pieris brassicae	68.5	Meghalaya
Crocidolomia binotalis	28.09-50.88	Karnataka
Cabbage borer (*Hellula undalis*)	30-58	Karnataka
Chinese cabbage		
Sawfly (*Athalia proxima*)	36.5	Uttar Pradesh
Aphid (*Lipaphis erysimi*)	36.5	Uttar Pradesh
Chillies		
Thrips (*Scirtothrips dorsalis*)	11.8	Assam
	50	Tamil Nadu
	>90	Karnataka
Mite (*Polyphagotarsonemus latus*)	34.14	Andhra Pradesh
Brinjal		
Fruit and shoot borer (*Leucinodes arbonalis*)	63	Haryana
	50	Tamil Nadu
	48	Maharashtra
	54-66	Karnataka
Tomato		
Fruit borer (*Helicoverpa armigera*)	22.39-37.79	Karnataka
Okra		
Fruit borer (*H. armigera*)	22.1	Himachal Pradesh
Jassids (*Amrasca devastance*)	54-66	Karnataka
White flies (*Bemisia tabaci*)	54.04	Rajasthan
Shoot and fruit borer (*Earias vitella*)	54.04	Rajasthan
	22.79-37.52	Punjab
	38.43	Uttar Pradesh
Cucurbits		
Fruitfly (*Bactrocera cucurbitae*)		
Bittergourd	60-80	Uttar Pradesh
Cucumber	20-39	Assam
	80	Himachal Pradesh
Bottle gourd	63	Gujrat
Muskmelon	76-100	Rajasthan
Snake gourd	63	Assam
Sponge gourd	50	Andhra Pradesh

caused due to increased awareness among the end users and concerns about the deteriorating ecological situations among the eco-campaigners. However, it must be endured in our minds that there is no possible one hundred percent alternatives to chemical pesticides and they must be applied wherever situations arises. By incorporating more botanical means in existing integrated pest management the bio-intensive pest management, practices for major vegetable crops becomes more sustainable which is discussed in Tables 9.1 and 9.2.

Malvaceous Crops

Okra (*Abelmoschus esculantus* L.) (Moench) is the major economically important vegetable crop belongs under this family. This crop is being attack by various pests but major pests are leaf hopper *Amrasca bigutila bigutula* (Ishida), white flies *Bemisia tabaci* (Gennadius) which is also the vector of vein clearing disease, the next goes the fruit borer *Earias insulana* (Boisduval), *Earias vitella* (Fabricius). The tolerant/less susceptible varieties are the key component of IPM programme. They know to reduce the pest occurrence by non-preference, antibiosis and/or tolerance. Varieties that are tolerant to some important pests are MP-7, AE-57 Punjab-7, Clemson Spineless, for E. vitella; New Selection, Punjab Padmini against *Amrasca kerri*; Parbhani Kranti, Varsha Uphar, Arka Anamika, IC-7194, and IC-13999 for Yellow Vein Mosaic (Annon, 2001). Conservation of various natural enemies like parasitoids and predators is very important, the parasitoids like *Erythmelus empoascae, Stethymium empoascae,* are known to curb down the *Amrasca bigutula bigutula* while *Bracon hebetor* (Say), *Rogas testaceus, Microbracon lefroyi* are found effective against the borers *Earias* species. The *Aphelinus flavipes* parasitizes *Aphis gossypii* and *Eurytoma spp.* parasitizes Myzus hibisci. Apart from these parasitoids, the generalized predators like coccinellids, syrphids, spiders, staphylinids, predatory mites and anthocorid bugs, which are so dominant in the okra ecosystem, which need to be conserved, as they are highly susceptible to the applied pesticides. The microbial controls are one of the impeding forces to pest. Application of white muscardine fungus *Beauveria bassiana* @ 1 x 10^4 cfu/g two times with a week interval adequately managed whitefly nymphs and adult. Commercial formulations of *Bacillus thuringiensis var. kurstaki viz.* Biolep followed by Bioasp known to be most effective at various concentrations against *E. vitella* (Pandy *et al.*, 1999). The best part of botanical insecticides that they do not exert any adverse effect on the natural beneficial predatory fauna such as coccinellids, syrphids, and spiders and hence found suitable in okra IPM. They are tested extensively against these pests. Leaf extract of *Parthenium hysterophorus* 3 per cent was effective against leaf hopper. However, it was at par with rhizome extract of *Acorus calamus* at 3 per cent, leaf extract of *Ocimum basilicum* 3 per cent and *Datura strumarium* 3 per cent (Pawar *et al.*, 2001). Nicotine sulphate similarly found effective in curbing leaf hopper population. The combination of commercial formulation of *Bacillus thuringiensis* (dipel) with lower with some lower concentration of insecticides like endosulfan, fenvalerate, multineem carbaryl, acephate, and dipel alone were tested against okra shoot and fruit borer *Earias vitella* Fab. The dipel + endosulfan and dipel + fenvalerate were found very effective in reducing per cent shoot and fruit infestation (Tomar 1998).

Solanaceous Crops

The Solanaceous family in the vegetable group is the major and economically important that include tomato, chillies (*Capsicum annum*) and brinjal (*Solanum melongena* L.), widely cultivated in Indian subcontinent for their taste and nutritive value.

Tomato

Tomato (*Lycopersicon esculentum* Mill.) is a very common constituent of Indian diet. It is attack by an array of pests, however the major damage is caused by fruit borer (*Helicoverpa armigera* Hubner), serpentine leaf miner (*L. trifoli,*) white flies, (*B. tabaci*) and thrips (*Frankliniella occidentalis* Pergande), the latter two being the vector for viral diseases. Resistant varieties available are Roma AC and Punjab Kesari against fruit borer while for sucking pests we have Pusa Ruby, Pusa Sheetal and Pusa Gaurav, and also high yielding (Singh 1999). Other measures of control like by using marigold as a trap crop under cultural control is getting popularity among the farmers, here a row of marigold *Tagetas spp.* after every 14 rows of tomato is sown and should be 15 days older than tomato plants. The fruit borer tends to lay maximum egg on marigold flowers, from where larval movement is insignificant. Monitoring need to be done on the top three leaves. Flowers can be collected and destroyed (Srinivasan *et al.*, 1993) the egg parasitoids *Trichogramma pretiosum* (Riley) @ 1.0 lakh ha^{-1} six times from flower initiation stage at weekly intervals causes' significant drop in fruit damage (Krishnamoorti and Mani 1988). From predatory fauna fourteen species of predatory insects belonging to four families under three orders were found to attack whiteflies (*Aleurodicus disperses*). Among them a nitiduid predator, *Cybocephalus spp.* (new record in India) and coccinellid predators *viz., Axinoscymnus puttarudriahi, Cryptolaemus montrouzieri* (*Mulsant*), *Chilocorus nigrita* (*Nigrita*) and *Scymnus coccivora* were abundant. Among the chrysopids, Mallada astur and *Apertochrysa spp.* found abundant (Geetha *et al.*, 2002). India is rich in tree flora which possesses wide range of bioactive compounds like tannin and flavonoids. Various botanicals which derived from the neem and other plant species are conventionally utilized in Indian pest control methods. They are not only popular but also very effective in certain conditions as well as most compatible under IPM most popular being the Indian neem, *Azadirachta indica* (*A. Juss.*). Neem has diverse biological effects on insects. It has antifeedant, ovicidal deterrent, repellent, insect growth regulator, sterilant, mating disrupter and toxic properties (Singh and Doharey 2001). The use of neem and pongam (*Pongamia pinnata* L.) cakes shows satisfactory results for the management of tomato pests (Krishna Kumar and Krishnamoorthy 2002). In South India neem extract for the management of Tobacco caterpillar (*Spodoptera litura* Fabricius) is so popular, which is not only cheaper but also as effective as various synthetic insecticides (Joshi *et al.,* 1993). A study conducted with combination of *Bt* + *Ha*NPV + neem oil 0.3 per cent recorded the highest yield increase over the control plots and noted higher cost benefit ratio. It was witnessed that BIPM plots of tomato recorded higher yield as compared to farmer practice plots (Praveen and Dhandapani 2001).

Brinjal

Brinjal (*S. melongena*) a special member of solanaceous family in vegetables as it gives better returns over investment to the farmers as well as bank of nearly all kinds of minerals and trace elements, yet the yield of this crop get lessen by numerous insect pests. The major pests include shoot and fruit borer (*Leucinodes orbonalis* Guenee), Hadda beetle (*Epilachna viginctioctopunctata* Fabricius), aphids (*Aphis gossypii* Glover), ash weevil (*Myllocerus subfasciatus*). Cultural operations are so effective against soil borne nematodes. Trap cropping with sweet potato (*Ipomea batata* Sheela *et al.,* 2002), and marigold (*Tagetus erecta*) are effective in controlling *Melidogyne incognita*. Brinjal Sree bhadra rotation brings 47 per cent reduction in nematode population. Inter cropping with single and double rows of coriander (*Coriandrum sativum* L.) with border crop of brinjal significantly reduced the incidence of brinjal shoot and fruit borer. The results shown that brinjal + two rows of coriander intercropping system was so effective in curbing down the shoot damage ranging from 49.85 to 98.76 per cent (David 2000). The clipping of infested shoots in brinjal also helps to attain less damage (Chiranjeevi *et al.,* 2002). Several workers reported different pivotal groups of natural enemies on *L. orbonalis* (Ayyar, 1927), *Eriborus argentiopilosus* (Tewari and Singh 1987), *Bracon chinensis* (Nair 1967), *Apanteles* spp., *Chelonus* spp (Gokulpur 1967 and Navasero and Calilung 1990) and the chalcids *Brachymeria obscurata* (Walker). Role of *Pediobius foveolatus* (J.C. Crawford) in the biocontrol of *H. viginctioctopunctata* was studied (Jayanthi and Alexander 2000). Investigations revealed that the third and fourth instar grub stages of the host are most preferred and susceptible for parasitisation than other stages. From the various microbials, baculovirus infecting *L. orbonalis* under field conditions is reported to be effective and needs further exploitation and experimentation (Tewari and Sandana 1987). Several other botanicals were also found to be effective against pest complex of brinjal. Results on antifeedant properties of two commercial neem formulations *viz.,* Neem Azal T/S and Neem Azal F in comparison with standard insecticide endosulfan 35 EC against brinjal sphingid larvae, leaf beetle grubs and ash weevil adults revealed that Neem Azal F 1 ml/l was the most potent (Kumar and Babu1999). Plant extracts of *Annona squamosa, Argemone americana* and *Calotropis giganteca* brought considerable reduction of *Henosepilachna viginctioctopunctata* population (Banerji *et al.,* 1985) borer incidence can bring down with help of four times of neem cake applications during the crop growth phase. The yield was improved by 48 per cent as compared to control (Sreenivasamurthy *et al.,* 2002). Ash weevil has recently emerged as a serious pest in major brinjal growing tracts. It munches on the leaf margins while the grubs damage the roots leading to withering of leaves and caused impact on photosynthesis. In case of severe attack the defoliation may also occur. During transplantation the application of neem cake @ 500 – 1000 kg/ha or drenching of neem seed kernel extract 4 per cent efficiently reduced the ash weevil incidence (Anon 2001).

Chillies

In India chilli (*C. annuum*) is grown extensively. It is key constituent of Indian vegetarian and non-vegetarian cuisine. It is desired for pungency and recipes without it are quite difficult to imagine. The major pests of chilli include *thrips, Scirtothrips* dorsalis (Hood); mites, *Polyphagotarsonemus latus* (Banks); fruit borer, *S. litura* and *H. armigera*. There are some varieties which are comparatively tolerant or less susceptible *viz.* HC 144 against whitefly and aphid and G5, K2 and X 235 against thrips (Table 4). Various natural enemies like *Encarsia haitiensis* and *E. guadeloupae*, can suitably control the major pests of chillies. *E. guadeloupae* caused 20.54 per cent parasitism in Andhra Pradesh and 70.26 per cent in Kerala (Mani *et al.*, 2002). Couple of releases of *T. chilonis* @ 2.5 lakh per hactare caused heavy egg reduction of (55.79 per cent) *H armigera* (Kulkarni and Shekharappa 2001). During the time of transplanting, seedling root dipping in imidacloprid (0.2 per cent) solution three hours prior to planting, followed by three sprays of imidacloprid (0.1 per cent) recorded the least incidence of chilli mosaic disease which is vectored by insects like whiteflies (Thiruvudainambi *et al.*, 2002). Thiacloprid (Calypso 240 SC), from chloronicotinyl group systemic insecticide 225 and 300 ml/h applied twice four and seven weeks after transplanting resulted in effective control of thrips up to 9 weeks after first spraying and expressively enhance the fruit yield. Residues were below measureable level in all treatments of thiacloprid in green, ripe and dry chillies (Dhandapani *et al.*, 2002). Newer chemistry like fenpyroximate 30 g *a.i.* per ha and diafenthiuron 600g *a.i.* per ha brought down the eggs and the active stages of *P. latus* by 90-95 per cent followed by fenazaquin 125 g *a.i.* per ha. Bifenthrin and fenpropathrin were less active against *P. latus* (Srinivasa *et al.*, 2002)

Cruciferous Crops

The cruciferous groups also known as Cole crops, which includes economically important crops like Cabbage (*Brassica oleracea var. capitata*), Cauliflower (*B. oleracea var. botrytis*), Knol khol (*B. caulorapa* L.) all over the world. However poor yield is mainly attributed to the incidence of insects pest predominantly diamondback moth *i.e.* (DBM) *Plutella xylostella* (L.), leaf webber (*Crocidolomia binotalis* Zeller), cabbage web worm (*Hellela undalis* Fab.), head borer (*S. litura*) and aphid (*Brevicoryne brassicae* L.) infesting the crop in various seasons. For stem borer for cauliflower the tolerant/less susceptible varieties are Early Patna, and Kathmandu local. Intercropping of one row of tomato (30 days after cabbage) one row of cabbage found promising to manage the DBM and leaf webber incidence (Srinivasan and Veeresh 1986). Similarly the practice of trap/intercropping of mustard in cruciferous vegetables reported to be so helpful in lowering the pest status of DBM (Arora *et al.*, 2000). In undisturbed fields of cruciferous crops there are plentiful numbers of biocontrol agents, which can take care of the pest status but our unnecessary interference with chemical pesticides has reduced their population, hence it is necessary to introduce and augment their population. Introduction of parasitoid *Apenteles plutellae* (Kurdyumov) 1000 adults' release every two week interval up to the harvest helps to curb down the DBM population. Utilization of pentatomid predator, *Eocanthecona furcellata* (Wolf) was found very effective as they feed voraciously on *Crocidolomia* and *Hellula spp* (Annon 1995). Microbials derived from actinomycetes, bacteria, fungi, and viruses are found to be very effective against these pests (Annon 1995). *Metarhizium anisopliae* var *anisopliae* (Gopalakrishnan *et al.*, 2002) *Spinosyns Spinosad* 45 SC were effective against *P. xylostella, S. littoralis* and *H. undalis* (Dey *et al.*, 2001). Vertimec 1.8 EC, which is extracted from naturally occurring soil microorganism *Streptomyces avermetilis* was found very effective against DBM, (Murugan and Ramachandran 2000). *Bt* based formulations are dominating the pest control operations in cruciferous crops. Various research workers reported their efficacy against the lepidopteran pests (Malathi *et al.*, 1999). The utilization of overhead sprinkler system aids in establishing the larval parasitoid, *A. plutellae* for control of DBM (Tabashnik 1956). In order to avoid the buildup of resistance to insecticides in DBM and to conserve it's natural enemies, the selective use of avermectin, *Bt*, a mixture of virus and *Bt*, selective insecticides like diafenthiuron and chlorfluazuron has been proposed (Zuhua and Shusheng 1998).

Cucurbitaceous Crops

Cucurbits are an important group of vegetables popularly called as gourds, they are served as raw and cooked which includes Pumpkin (*Cucurbita moschata* Poir), Ash gourd (*Benincasa hispida* Cogn), Bitter gourd (*Momordica charantia* L.), Snake gourd (*Trichosanthes cucumeria* L.), Ridge gourd (*Luffa acutangula* Roxb), Cucumber (*Cucumis sativus*) *etc.* They are suitable for cultivation round the year. This group is being attacked by an array of insect pests which not only feeds the crop but also affect the appearance of the marketable yield. These pests include red pumpkin beetle (*Aulacophora foveicollis* Lucas), fruity fly (*Dacus cucurbitae* Coquillet), leaf miner (*L. trifolii*). Parasitoids like Hymenoptera (*Hemiptarsenus varicornis*) an endolarval parasiotid of leaf miner (Krishnamoorthy and Krishna Kumar 2002) is found to be very effective against this pest. Predators like *C. carnea, Opius compensatus, O. incisus, Spaladia philippinensis* are effective against fruit fly (Annon 2001). Unlike other crops cultural control is proved to be equally effective as additional control options and very economical, and requires very little inputs. Erection of as much as of twenty yellow sticky traps smeared with castor oil as adhesive works effectively against whitefly *Aleurodicus disperses* (Geetha *et al.*, 2002), wrapping the fruits with paper bags protects against fruit fly

(Regunath and Gokulapalam 1999). Hand picking and destruction of pumpkin beetle and grubs from plant and stirring the soil eventually leads mortality of pupal stages of *D. cucurbitae*, erection of traps containing methyl eugenol (1 ml/l) + Vinegar/gur + Malathion (1 ml/l) and pulverizing the soil for controlling *D. cucurbitae* have been effective. Microbial treatments like *B. bassiana* @ 2 x10^6 cfu/g proved to be effective against pumkin beetle (Anon 2001). Various reports also suggest that nematode species *S. feltiae* effectively curbed these three species of leaf miners. The serpentine leaf miner, due to internal feeding habits is not exposed to outer environment and natural enemies. Studies indicated that neem seed kernel extract @ 5 per cent concentration found promising in control of leaf miner and recorded highest yield of bitter gourd. Karanj (*P. pinnata*) plant oil (29.68 per cent) was found next in the order of efficacy, Whereas Monocrotophos failed to offer satisfactory control of leaf miner (Rosaiah 2001). The available tolerant/less susceptible varieties for Fruit fly (*D. cucurbitae*) are MM22-2, Lakhazda, Khara, IHR 40 and 47 similarly for muskmelon; HISSAR II and Ghote in Bitter gourd; Arka Tinda in round gourds5.

Onion

In India without onion no curry or cuisine can be imagined it's a key component of nearly all dishes, and being attacked by various pests and lethal diseases. The major pests include onion thrips *Thrips tabaci* and onion fly *Delia antiqua*. Some of the tolerant/less susceptible varieties for thrips are Pusa Ratnar, Arka Niketan, PBR-2, PBR-5, PBR-661. Still the natural enemies of onion fields are not well accounted, however generalized predators like coccinellids, syrphids, spiders and *C. carnea* 2 nos per plant against the sucking pests were found to be effective. Pollen cum nectar bearing flowering plants if planted on the bunds, can provide supplementary food to predators like chrysopids, coccinellids, predatory mites and parasitoids. Jet sprinkler of water via jet nozzles can be used to prevent thrips multiplication (Singh 1997) the ability of thrips to disperse contributes greatly to their pest status. The barrier cropping comprising of two rows of maize and outer row of maize and inner row of wheat effectively blocked the adult thrips movement. Onions with 2 lines of maize noted the least number of nymphs followed by outer line of maize and inner line of wheat (Anon 2001).

Future Thrust

Our Indian agriculture is beholding a revolution. We have our first transgenic crop as cotton, and still some transgenic vegetable crops are in the waiting list to be allowed to be cultivated and marketed like tomato (Cry1Ab) to be developed against fruit borer; its glasshouse trials are over. Brinjal (Cry1Ab) is being transformed to combat *Leucinodes* problem; its glasshouse experiments are going on, similarly for cauliflower and cabbage (Cry 1H and Cry 9c) are getting developed to combat lepidopterous pests and its glasshouse experiments are under trail. In spite of transgenic crops our future research should include how to reduce insecticide uses and their residues from vegetables by adopting biodegradable and ecofriendly IPM technology *viz.*, trap and mixed cropping and avoidance of peak activity period of pest by altering sowing and harvesting time these techniques do not require inputs but slight medication of cropping pattern and alternation of microclimate sometimes leads to breakdown of severe pests. Farmer's participation and awareness is a key factor in making any IPM a success story at field level, case study, indigenous knowledge, training and development of resistant varieties as well as pest dynamics *etc.* Studies on the importance of natural enemies their mortality factors and the role-played by each factor at different locations in different seasons can be a pivotal factor in an IPM strategies and decision making. Need based research like ovipositional deterrent for fruit fly (*D. curcubitae*) have to be intensified? In cruciferous crops detailed studies are required on the bioecology and methods of mass production of important natural enemies like *A. plutellae,* as well as natural enemies in onion crop ecosystem need to study and documented in proper way. This work should be undertaken urgently so as to prevent the further loss of ecology and to continue the sustainable farming.

CONCLUSION

The assessment of the available literature express that there are several numbers of ecofriendly ways available to lessen down the pesticide exploitation in vegetable cultivation and production optimization. There is a necessity to recognize the potential of indigenous biocontrol agents and care should be given to conserve them at field level itself. Importing the exotic parasitoids and predators is also needed to be step up for several crop pests to strengthen the natural enemy fauna. Survey and documentations of various natural enemies and conservation methods for parasitoids and predators as well as use of pheromones should be increased.

REFERENCES

Annon, 2001. Integrated pest management package for Cucurbitaceous crops (Eds.) Sipra Bhattarcharya and Mamta Madhi. Offset Press, Directorate of Extension, Ministry of Agriculture, Government of India, New Delhi, India.

Annon, 2001. Integrated pest management package for okra (Eds.) Sipra Bhattarcharya and Mamta Madhi. Offset Press, Directorate of Exten- sion, Ministry of Agriculture, Government of India, New Delhi, India.

Annon, 2001. Yearly report of Asian Vegetable Research and Development (AVRDC).1995 Management of insects and plant diseases. Taiwan, 109-116 pp.

Anon, 2001. Integrated pest management package for cucurbitaeceous crops (Eds.) Sipra Bhattarcharya and Mamta Madhi. Offset Press, Directorate of Extension, Ministry of Agriculture, Government of India, New Delhi, India.

Anon, 2001. Pest management. In: Handbook of horticulture. (Ed.) Chadha, K.L. ICAR, New Delhi. 785-799 pp.

Anon, 2001. Yearly report of Asian Vegetable Research and Development (AVRDC). 1995. Management of insects and plant diseases. Taiwan, 109-116 pp.

Anon, 2001.Integrated pest management package for onion (Eds.) Sipra Bhattarcharya and Mamta Madhi. Offset Press, Directorate of Extension, Ministry of Agriculture, Government of India, New Delhi, India.

Arora R., Battu, G.S. and Bath, D.S. 2000. Management of insect pests of cauliflower with biopesticides. *Indian J. Ecol.* **27**(2): 156-162.

Ayyar, T.V.R. 1927. The parasitic hymenoptera of economic importance noted from South India. Bull. *Entomol.* Res. **18**: 73-78.

Banerji, R., Misra, G. and Nigam, S.K. 1985. Role of indigenous plant material in pest management. *Pesticides,* 32-37 pp.

Chiranjeevi, C., Reddy, I.P. Narayanamma, M. and Neeraja, G.2002. Bioefficacy of newer acaricidal molecules against chilli mite, *Polyphagotarsonemus latus* Banks (Acari: Tarsonemidae). In: International Vegetable Conference. Nov. 11-14, Bangalore, India, 265 p.

Chiranjeevi, C., Reddy, I.P. Narayanamma, M. and Neeraja, G.2002. Effect of shoot clipping and insecticides on the incidence of fruit borer in brinjal. In: International Conference on Vegetables. Nov. 11-14, Bangalore, India. 271 p.

David Abilash, R.B., 2000. Studies on development of IPM techniques for brinjal shoot and fruit borer *L. orbonalis. M.Sc. thesis,* Tamil Nadu Agricultural University, Coimbatore, India. 1-67 p.

Dey, P.K. and Somchoudhury, A.K., 2001. Evaluation of spinosyns A+D (Spinosad 48 SC) against lepidopteran pest complex of cabbage and its effect on natural enemies of diamondback moth under field conditions of W.B. Pestology, **25**(1): 54-58.

Dhandapani, N., Dhivahar, P. and Murugan, M., 2002. Thiacloprid – A new molecule for the management of thrips of chillies. In: International vegetable conference, Nov.11-14, Bangalore, India. 259 p.

Geetha, B., Swamiappan, M. and Kempuchetty, N. 2002. Predators for the biological control of the spiralling whitefly *Aleurodicus dispersus* on vegetables. In: International Conference on Vegetables. Nov.11-14. Bangalore, India. 261 p.

Geetha, B., Swamiappan, M. and Kempuchetty, N. 2002. Predators for the biological control of the spiraling whitefly Aleurodicus dispersus on vegetables. In: International Conference on Vegetables. Nov.11-14. Bangalore, India.261 p.

Gokulpur, R.S., 1970. Records of braconids and ichneumonids from Central India. *Indian J. Ent.* **32**(4): 383-385.

Gopalakrishnan, G., Krishnamoorthy, A. and Gangavisalakshi., P.N.2002. *Metarhizium anisopliae* var. *anisopliae*: A potential entomopathogenic fungus for the management of *P. xylostella* L. on cabbage. In: International Conference on Vegetables. Nov. 11-14, Bangalore, India, 262 p.

Jayanthi, S. and Alexander Jessudasan, R.W.2000. Role of *Pedobius foveolatus* (Hymenoptera: Eulophidae) in the biocontrol of the brinjal spotted leaf beetle: *Henosepilachna viginctiotopunctata* (Coleoptera: Coccinellidae). *Pestology*, **24**(5): 33-36.

Joia, B.S., Gursharan Singh, Navdeep Singh and Darshan Singh. 2001. Field evaluation of cartap hydrochloride for the control of *Plutella xylostella* (L.) on cauliflower. In: Proceedings of the second national symposium on integrated pest management in horticultural crops: New molecules, biopesticides and environment. Oct. 7-19, Bangalore, India.

Joshi, B.G., Sriprakash, K. and Murthy, P.N.S. 1993. Efficacy of neem seed kernel suspension, neem seed bitter, pongamia cake water extract, *Bacillus thuringiensis,* Berliner, nuclear polyhedrosis virus alone and in combination against tobacco caterpillar, *Spodoptera litura* F. in nurseries. In: Chari, M.S. and Ramaprasad, G. (Eds), Botanical pesticides in integrated pest management. Proc. Symposium on botanical pesticides in IPM (Rajmundry, India). 154-158 pp.

Krishna Kumar, N.K. and Krishnamoorthy, P.N. 2002. Advances in the use of botanicals for the IPM of major vegetable pests. In: International Conference on Vegetables. Nov. 11-14. Bangalore, India. 233 p.

Krishnamoorthy, P.N.and Krishna Kumar, N.K 2002. Advances in the use of botanicals for the IPM of major vegetable pests. In: International Conference on Vegetables. Nov. 11-14. Bangalore, India. 233 p.

Krishnamoorti, A. and Mani, M. 1988. Feasibility of managing *H. armigera* on tomato through parasitoid. In: National Workshop on Heliothis management. Feb. 18-19. TNAU, Coimbatore.

Kulkarni, K.A. and Shekharappa, 2001. Integrated management of chilli fruit borer, *H.armigera*. In: Proceedings of the second national sympo- sium on integrated pest management in horticultural crops: New molecules, biopesticides and environment. October 17-19, Bangalore, India, 59-60 p.

Kumar, S. P. and Babu. S. 1999. Evaluation of antifeedant properties of neem azal against certain leaf feeders of brinjal. Pestology, **22**(10): 62-64. *L. orbonalis* Guen. Curr. Sci., **56**(13): 671-672.

Malathi, S., Sriramulu, M. and R. Babu, T.1999. Evaluation of certain ecofriendly insecticides against lepidopteraous pests of cabbage. *Indian J. Ent.* **61**(2): 127-133.

Mani, M., Dinesh, M.S.and Krishnamoorthy, A. 2002. Survey and evaluation of natural enemies of the exotic spiralling whitefly, *Aleurodicus dispersus* on vegetables in peninsular India. In: International Conference on Vegetables. Nov. 11-14, Bangalore, India, 238 p.

Mukharjii, K.G. and Rajak, R.K. Aditya books private Ltd. New Delhi, India.

Murugan, M. and Ramachandran, K. 2000. Bioefficacy of vertimec 1.8 EC against diamond back moth, *Plutella xylostella* (L.) on cabbage. *Pestology*, **24**(1): 12-14.

Nair, R.K., 1967. *Leucinodes orbonalis* as a serious pest of potato at two places in Mysore state. *Indian J. Ent.* **29**(1): 96-97.

Navasero, M.V. and V.J. Calilung. 1990. Biology of the egg plant fruit and shoot borer, *L. orbonalis* in Philippines. *Philippine Entomologist*, **8**(2): 749-760.

Pandy, N.K., Ojha, P.K. and Upadhyay, K.D. 1999. Studies on the bioefficacy of *B. thuringiensis* formulations against *E. vitella* Fabr. under laboratory conditions. RAU. *J. Res.* **9**(1): 57-61.

Paroda, R.S. 1999. For a food secure future. The Hindu survey of Indian Agriculture.

Pawar, D.B., Warade, S.D. and More, T.A. 2001. Efficacy of different plant extracts against leaf hopper (*Amrasca devastans* Dist.) and fruit borer (*E. vitella*) of Okra. In: Proceedings of the second National Symposium on Integrated Pest Management (IPM) in Horticultural Crops: New Molecules, Biopesticides and Environment. October 17-19. Bangalore, India.66 p.

Praveen, P.M. and Dhandapani, N. 2001. Ecofriendly management of major pests of tomato *Lycopersicon esculantum* (Mill.) in India. *South Indian Horti.* **49** (special): 261-264.

Praveen, P.M., Dhandapani N. and Kennedy, J.S. 2001. Efficacy of *Bacillus thuringiensis* (Berliner) formulations for the management of tomato fruit borer, *Helicoverpa armigera* (Hubner).Pestology, **25**: 58-61.

Regunath, P. and Gokulapalam, C.1999. IPDM in bittergourd experiences in participatory approaches. In: Proceedings of Hort. India 277-279.

Rosaiah, B. 2001. Effect of different plant products against serpentine leaf miner Liriomyza trifolii on ridge gourd (*Luffa accutangula*). *Pestology*, 25(4): 7-9.

Sardana, H.R. 2002. Integrated pest management in vegetables. In: Training manual -2, Training on IPM for Zonal Agricultural Research Stations. May21- 26.105-118 pp.

Sheela, M.S., Nisha, M.S. and Mohandas, C., 2002. Impact of trap cropping and crop rotation for the management of root-knot nematode in vegetables. In: International Conference on Vegetables. Nov. 11-14. Bangalore, India. 245 p.

Shivalingaswamy, T.M., Satpathy, S., Singh, B. and Akhilesh Kumar. 2002. Predatorprey interaction between jassid *Ammrasca devastance* and staphylinid in Okra. In: International Conference on Vegetables. Nov. 11-14. Bangalore, India. 275 p.

Shivalingswami, T.M., Satpathy, S. and Banergee, M.K. 2002. Estimation of crops losses due to Insect pests in vegetables Ed. by Sarath Babu, B., Varaprasad, K.S. Anitha, K., Prasada Rao, R.D.V.J., Chakrabarthy S.K. and Chandurkar, P.S. Resource management in plant protection.1: 24-31.

Singh Dilbag., 1991. Pests of solanaceous vegetables. In: A decade of research on pests of host-crops 1980-90. Group discussion of entomologists working in coordinated projects of Hort crops held at CIHNP, Lucknow Jan. 28-29.

Singh, R. P. and Doharey, K.L. 2001. Integrated pest management in vegetables. In: Training manual -2, Training on IPM for Zonal Agricultural Research Stations. May 21-26.105-118 pp.

Singh, U.C., Singh, R. and Nagaich, K.N. 1999. Evaluation of tomato varieties against jassid (*E. devastance*) white fly (*B. tabaci*) and leaf curl. *Indian J. Ent.* **61**(2): 173-176.

Sreenivasamurthy, D., P.N. Krishna Moorthy, M. E. Raja and Shivanna. 2002. Economic assessment of pest management in brinjal and cabbage using neem products. In: International Conference on Vegetables. Nov. 11-14, Bangalore, India. 243 p.

Srinivas, P.S. and K.E. Lawande. 2002. Barrier cropping a new method for the management of thrips in onion. In: International conference on vegetables. Nov. 11-14, Bangalore, India. 237 p.

Srinivasan, K. 1993. Pests of vegetable crops and their control. In: Advances in Horticulture, 6: Vegetable crops Eds. K.L. Chadha and G. Kalloo. Malhotra Pub. House, New Delhi. p. 859-886.

Srinivasan, K. and Krishnamoorthy, P.N.1991. Indian mustard as a trap crop for management of major lepidopterous pests on cabbage. In: Trop. Pest Management, **37**: 26-32.

Srinivasan, K., Krishnamoorthy, P.N. and Prasad, R. 1993. Evaluation of different trap crops for the management of fruit borer, *H. armigera* on tomato. In: Abs. Golden Jubillee Symp. Horticultural Research: Changing Scenario. May 24-28. Bangalore, India. 259 p.

Srinivasan, K. and Veeresh, K.1986. The development and comparison of visual damage threshold for chemical control of *Plutella xylostella* and *Crocidolomia binotalis* on cabbage in India. *Insect Sci.* Appli. **7:** 547- 557.

Tabashnik, B.E., 1986. Model for managing resistance to fenvalerate in the diamondback moth (Lepidoptea: Plutellidae), in *J. Econ. Entomol.* **80**: 1091-1099.

Tewari, G.C. and Sandana, H.R.1987. *Eriborus argentiopilosus* (Caweras) a new parasite of *L. orbonalis.* Entomon. **12**(3): 227-228.

Thiruvudainambi, S., S. Doraisamy, D. Veeraraghavanatham and K. Srinivasan. 2002. Chemical and non-chemical methods of managing mosaic disease of chilli. In: International conference on Vegetables, Nov. 11-14, Bangalore, India, 258 p.

Tomar, R.K.S. 1998. Efficacy and economics of biopesticides and insecticides combinations against okra shoot and fruit borer. Indian J. Ent. **60** (1): 25-28

Zuhua, S. and Shusheng, L.1998. Toxicity of insecticides commonly used in vegetable fields to the diamondback moth, *Plutella xylostella* and its parasite, *Cotesia plutellae,* in Chinese *J. Biol. Control.* **14**(2): 53-57.

Chapter 10

Non-Chemical Management Strategies of Soil Borne Plant Pathogens of Vegetables

Abhishek Singh & Sachin Kumar Jain

Department of Plant Pathology, Amar Singh (PG) College, Lakhaoti, Bulandshahar, Uttar Pradesh

ABSTRACT

Fungi and nematodes-caused soil-borne diseases are key yield-limiting factors that are difficult to control. Plant parasitic nematodes alone have been estimated to take away ten percent of worldwide agricultural output, resulting in annual economic losses of more than $125 billion. Soil borne pathogens show broad host range and excellent survival ability that make it difficult to manage. Fusarium spp., Verticillium spp, Rhizoctonia solani, Meloidogyne spp., Phytophthora infestans and Alternaria spp etc. were identified as the most frequent fungi and nematodes responsible for a variety of soil-borne diseases. In this chapter of the book we will try to explore chemical free management strategies for soil borne plant pathogens. Various practices i.e. field sanitation, using resistant cultivars, crop rotation, soil amendments, manures, soil solariation, biofumigants, seaweeds, animal manures and microbial management can be used to management of soil borne diseases.

Keywords: *Soil inhabitant, Soil invader, Seaweeds, Biofumigants, Trichoderma.*

INTRODUCTION

Soil borne phytopathogens can reduce the yield of vegetable crops. These pathogens are able to survive in the soil for many years. Many diseases caused by soil borne pathogens are difficult to predict, detect and diagnose. The term soil borne pathogens can be defined as pathogens that cause plant diseases via inoculums that come to the plant by the way of soil. Soil borne pathogens can be divided into soil inhabitant (those pathogens able to survive for relatively long time) and soil invader or soil transient (those pathogen able to survive for relatively time). 'soil inhabitants', have broad host ranges that include weeds and produce longlived survival structures. Important soil-borne pathogens include fungi, fungi-like organisms, bacteria as well as viruses and plant parasitic nematodes (Baysal-Gurel *et al.*, 2012). Fungal pathogens including species of *Fusarium*, *Rhizoctonia*, *Verticillium*, *Sclerotinia* and *Macrophomina phaseolina*, cause the loss of billions of dollars each year. Many soil-borne fungi persist in the soil for long periods by producing resistant survival structures such as chlamydospores, oospores and sclerotia (Baysal-Gurel *et al.*, 2012)

The term biological control refers to the purposeful utilization of introduced or resident living organisms other than disease resistant host plants to suppress the activities and populations of one or more plant pathogens (apsnet). Bio-priming is a new technique of seed treatment that integrates biological (inoculation of seed with beneficial organism to protect seed) and physiological aspects (seed hydration) of disease control. It is recently used as an alternative method for controlling many seed- and soil-borne pathogens. According to the USDA (United States Department of Agriculture), National Organic Standards Board (NOSB), organic agriculture is defined as "an ecological production management system that promotes and enhances biodiversity, biological cycles, and soil biological activity. It is based on the minimal use of off-farm inputs and on management practices that restore, maintain or enhance ecological harmony. The primary goal of organic agriculture is to optimize the health and productivity of interdependent communities of soil life, plants, animals and people".

Management Practices for Soil Borne Diseases

1. Field Sanitation

Soil borne plant pathogens can survive in the field with their resting spore like chlamydospores, microsclerotia, oospores or sclerotia for a long period of time even in absence of their host. Sanitation includes any sort of activities which are aimed to prevent the spread of pathogens by removing diseased and infected plant parts, decontamination of tools and equipment and washing hands. Tillage is also good sanitation measures which can expose infected plant materials to direct sun light. Tools that are used should be disinfected when moving equipment between different plants or fields using different methods such as heat treatment, ultraviolet (UV) treatment and chlorine treatment (Panth, 2020). These practices are especially important to prevent movement of soilborne pathogens such *Sclerotinia sclerotiorum*, *Phytophthora capsici*, *Verticillium dalhiae* and different species of *Fusarium*. Sanitation in the field needs to take place before, during and after the growing season. It helps to break the chain of disease cycle.

2. Resistant Cultivars

Use of disease resistant varieties for management of plant disease has been termed as painless method because it does not cost the farmer for anything. It also helps to maintain biological balance in the environment.

Table 10.1: List of some Resistant Varieties of Vegetables

Sl.No.	Crop	Name of Disease	Causal Organism	Resistant/Tolerant Variety
1.	Brinjal	Bacterial wilt	*Ralstonia solanacearum*	Swarna Shree, Swarna ManiSwarna Shyamli, Swarna Pratibha, Swarna Shobha Pusa Purple Cluster, Pusa Anupama, Arka Anand, Arka Keshav, Arka Nidhi, Arka Neelkanths,
		Bacterial wilt and Phomopsis blight (both)	*Ralstonia solanacearum* *Phomopsis vexan*	Swarna Abhilamb, Swarna shakti, Swarna mohit, Swarna ajay, Swarna neelima, JC-1, JC-2, Pant Samrat, Pant Brinjal Hybrid1, Azad B-2
2.	Tomato	Bacterial wilt	*Ralstonia solanacearum*	Arka Ananya, Arka Abhijit, Arka Abha, Arka Alok, Swarna Lalima, Swarna Naveen,
		Bacterial wilt and Early Blight	*Ralstonia solanacearum* *Alternaria solani*	Swarna Samridhi, Swarna Sampada, Swarna Vijaya, Swarna Deepti
		Root knot nematode	*Meloidogyne spp*	Pusa 120, Hisar Lalit, Pusa hybrid 2, Pusa Hybrid 4
3.	Chilli	Fusarium wilt	*Fusarium solani*	Phule Jyoti, Phule Mukta
4.	Cabbage	Black rot	*Xanthomonas campestris pv. campestris*	Pusa mukta
5.	Cauliflower	Black rot	*Xanthomonas campestris pv. campestris*	Pusa Shubra, Pusa Snowball K-1, Pusa Snowball Kt-25,
6.	Onion	Purple blotch	*Alternaria porri*	Arka Pitamber, TNAU hybrid CO2, Arka Kalyan
7.	Potato	Late blight	*Phytophthora infestans*	Kufri arun, Kufri himalina, Kufri chipsona-1, 3, 4, Kufri gaurav, Kufri garima, Kufri girdhari
		Late blight and Early blight	*Phytophthora infestans and Alternaria solani*	Kufri badshah, Kufri jeevan, Kufri khyati
		Late blight and wart disease	*Phytophthora infestans and Synchytrium endobioticum*	Kufri anand, Kufri chipsona -2, kufri frysona, Kufri giriraj, Kufri himsona, Kufri jawahar,

Sl.No.	Crop	Name of Disease	Causal Organism	Resistant/Tolerant Variety
		Early blight, late blight, wart disease	*Alternaria solani, Phytophthora infestans, Synchytrium ondobioticum*	Kufri jyoti, Kufri kanchan, Kufri kashigaro
		Wart disease, Late blight and charcoal rot	*Synchytrium endobioticum, Phytophthora infestans, Macrophomina phaseolina*	Kufri kumar
		Wart disease, Early blight and charcoal rot	*Synchytrium endobioticum, Alternaria solani, Macrophomina phaseolina*	Kufri chamtkar
8.	Spinach	Cercospora leaf spot	*Cercospora beticola*	Ooty (Sp) 1

Source: ICAR.

3. Role of Nutrients

Nutrients are important for growth and development of plants and they are important factors in disease control although plant disease resistance and tolerance (Agrios, 2005). Plant diseases are affected by the environment and especially by nutrient deficiencies and toxicities (Marschner, 1995; Krauss, 1999). The level of nutrients can affect the physiology and biochemistry and especially the integrity of the cell walls, membrane leakage and the chemical composition of the host, *e.g.* the concentration of phenolics can be affected by B deficiency (Graham and Webb, 1991).

The response to the N level was different in the facultative parasites, as when the plants were grown under high levels of N, they were more resistant to pathogens such as *B. cinerea* while in the case of obligate pathogens such as *Pseudomonas syringae pv. tomato, Ustilago maydis and Oidium lycopersicum* increased susceptibility was observed when plants were grown with high N supply (Hoffland *et al.*, 2000; Kostandi and Soliman, 1991).

The form of N is also important in plant diseases and the presence of notification inhibitors is important too (Huber and Graham, 1999; Celar, 2003; Harrison and Shew, 2001). At high NO_3 diseases are decreased in the case of *Fusarium oxysporum, Botrytis cinerea, Rhizoctonia solani* and *Pythium* spp. In contrast, at high NH4 disease is decreased in the case of *Pyricularia, Thielaviopsis basicola, Sclerotium rolfsii* and *Gibberella zeae*. The form of N can affect the pH of the soil and also the availability of other nutrients such as Mn. The level of N can affect the phenolics content of plants, which are precursors of lignin.

Nitrate-nitrogen decreased the severity of *Fusarium* wilt of Yellow Delaware chrysanthemum in comparison with half ammonium, half nitrate-nitrogen. Liming also reduced the severity of wilt. Nitrate-nitrogen together with calcium hydroxide decreased wilt additively. The severity of disease generally decreased with increasing soil pH (apsnet).

It has been shown that Mn soil applications reduce common scab of potato (Keinath and Loria, 1996) and *Sclerotinia sclerotiorum* in squash (Graham and Webb, 1991; Agrios, 2005). Boron has been shown to reduce diseases caused by *Plasmodiophora brassicae* in crucifers, *Fusarium solani* in bean, *Verticillium albo-atrum* in tomato and cotton, tobacco mosaic virus in bean, tomato yellow leaf curl virus in tomato (Dordas C., 2007). *Fusarium* wilt causes vascular wilt in vegetables is favored by warmer, low-pH soils. In contrast, application of nitrate fertilizers and application of lime, which reduces the availability of Mn and Fe, increases the pH and results in the reduction of the pathogen (Dordas C., 2007).

Table 10.2: Effect of Nitrogen in Disease Severity

Sl.No.	Pathogen	Low N Level	High N Level	References
1.	*Xanthomonas vesicatoria*	Increase	Decrease	Chase (1989)
2.	*Alternaria solani*	Increase	Decrease	Blachinski *et al.* (1996)
3.	*Fusarium oxysporum*	Increase	Decrease	Woltz and Engelhar (1973)

4. Soil Amendment and Manure

Organic manures made up of organic wastes, composts and peats have been proposed to control soilborne diseases and pests. *R. solani, Thielaviopsis basicola, V. dahliae,* species of *Fusarium, Phytophthora, Pythium* and *Sclerotium* are found to be managed effectively by the application of organic amendments (Bonanomi *et al.*, 2007; Shafique *et al.*, 2016). Organic amendments improve soil structure, boost water holding capacity and encourage the growth of other beneficial microbes that help to control soil borne pathogens. The increase in activity of microorganisms in the soil creates competition which may lead to effective suppression of harmful soil borne pathogens. Compost provides effective control of soil borne pathogens such as *Phoma* spp. and *Pythium* spp. The use of composted softwood and hardwood barks gave reproducible control of *Pythium ultimum* in lettuce, *R. solani* in cucumber, radish, and bedding plants under greenhouse conditions (Lagunas *et al.*, 1993; Stephens and Stebbins, 1985). The use of compost extract containing huge populations of micro-biota, *e.g.*, *Rhizobacteria, Trichoderma* and *Pseudomonas* species can enhance crop production and produce plant growth regulators and chemicals such as phenols or tannins which may have an antagonistic effect on soilborne pathogens (Welke, 2005) Biochar is a solid byproduct of biomass decomposition that can be utilised as an organic soil amendment and carbon source. Biochar's importance in the soil-rhizosphere ecosystem has already been demonstrated through studies on pH improvement, cation exchange capacity (Silber *et al.*, 2010) nutrient retention in light soils (Glaser *et al.*, 2002), and the promotion of beneficial microorganisms in the soil, all of which help to suppress soil borne pathogens.

5. Seaweeds

Application of seaweed has increased as an organic amendment basically for the purpose of fertilization of the crops but there is another side in the management of diseases that needs to be exploited. Seaweed contains growth regulators such as auxins, gibberellins and precursor of ethylene and betaine which affect plant growth. Seaweeds contain complex secondary metabolites that aid in the host's defense against diseases and parasites (Ara *et al.*, 2005). Seaweed may influence cell metabolism by inducing the formation of antioxidant compounds which may benefit plant development and stress tolerance (Zhang and Schmidt 2000). Anti-oxidant enzymes provide a degree of crop protection from free radical oxidants arising from normal metabolism and number of biotic and abiotic stresses. Betaines present in different extracts decreased nematode infestation (Wu *et al.*, 1997). Seaweeds also suppressed root rotting fungi and the root-knot nematode by producing antimicrobial compounds or synthesis of antioxidant molecules. Seaweeds contain 1-aminocyclopropane- 1-carboxylic acid (ACC) which has antimicrobial activity (Nelson and Van Standen 1985). Similarly, polyphenols are well known for their antioxidant activity and are widely distributed in seaweeds (Tariq *et al.*, 2011). A red seaweed (*Solieria robusta*) used as a soil amendment showed better suppression of root rotting fungus *F. solani* than Topsin-M (Sultana *et al.*, 2011a). Application of liquid concentrated formulation of brown algae *Ecklonia maxima* significantly reduced root-knot infestation and increased the growth of tomato plants.

6. Biofumigants

Crops of the family Brassicaceae (cabbage, broccoli, kale, turnip, radish, canola, cauliflower, rapeseed and various mustards) produce sulfur compounds called glucosinolates that release biologically active products during enzymatic hydrolysis. Isothiocyanates, the main metabolites of the glucosinolates hydrolysis are vapour compounds that are toxic soilborne pathogens. This process is referred to as biofumigation (Larkin and Honeycutt, 2006; Larkin and Griffin, 2007). Thus, soil amendment with Brassica napus seed meal reduced the incidence of apple root infection by *Rhizoctonia* spp. (Mazzola *et al.*, 2001; El-Sharouny, 2015). Xiao *et al.* (1998) observed that broccoli residues flail-mowed and allowed to dry on the soil surface for several days prior to incorporation were effective in reducing soil populations of *Verticillium dahliae* in cauliflower fields.

Plants belonging to the *Alliaceae* family, such as garlic, onion and leek also contain molecules with either a direct or indirect effect on pests and pathogens.

Crushed *Allium* spp. plants produce thio-sulfinates and related compounds such as disulfides which have the same spectrum of pesticide activity as methyl bromide (Auger *et al.*, 2004).

7. Soil Solarization

The mode of action of soil solarization is complex, involving direct thermal destruction of propagules, shifts in microbial populations. Solar heating involves the use of heat as a lethal agent for pest control through the use of clear plastic film to trap solar radiation and accumulate heat (Katan, 2000). It was reported that 5 days of solar heating were sufficient to eliminate 100 per cent of *Verticillium. dahliae* sclerotia at a depth of 5 cm. However, after 8 additional days of solarization results complete killing of the sclerotia was obtained at the 25 cm depth (Katan *et al.*, 1976). Similarly, after 19 days of solarization, mortality rates of *Sclerotium rolfsii* sclerotia at 5 and 20 cm depths were 100 per cent and 25 per cent, respectively, while, the respective 100 per cent and 80 per cent sclerotia mortality rates were observed after 21 additional days (Elad *et al.*, 1980).

8. Crop Rotation

Crop rotation is without a doubt a valuable method for plant disease management. However, it is insufficiently effective in reducing diseases caused by soilborne pathogens that have a wide host range or produce long-living survival structures, such as sclerotia, oospores or chlamidospores (Umaerus *et al.*, 1989)

Barley was highly recommended to reduce problems of pink root and Fusarium basal rot of onion (Schwartz, 2004). Oat-potato, annual ryegrass-potato or clover-potato crop sequences have been found to reduce and suppress *Rhizoctonia solani* inoculum levels in soil and subsequent disease development in a potato crop (Johnston *et al.*, 1994). Canola and rapeseed were in crop rotation for reducing the soilborne diseases commonly observed in those fields, including *Rhizoctonia* canker, black scurf and common scab with overall reductions ranging from 18 to 38 per cent. (Larkin *et al.*, 2010)

9. Animal Manures

Fertilization of animal manure is also an important component used to improve soil properties, plant health and yield. Stapleton *et al.* (2000) found that decomposition of animal manure releases ammonia (NH_3) which has an unfavourable effect on the survival or germination of some soilborne fungi and nematodes (Rodrguez-Kábana, 1986). Conn and Lazarovits (1999) showed that the application of fresh chicken manure can significantly reduce the incidence of potato scab, *Verticillium* wilt and parasitic nematodes. Swine manure was also effective in reducing potato scab. Combination of solarization and soil amended with chicken manure, chicken litter or plant residues was highly effective in controlling soilborne pathogens when compare to single treatment (Gamliel *et al.*, 2000; Stevens *et al.*, 2003).

Table 11.3: Some Important Biological Control Agents with their Mode of Action and Target Soilborne Pathogens

Sl.No.	*Biocontrol Agents*	*Target Pathogens*	*Mode of Action*	*References*
1.	*Bacillus* spp. (*B. subtilis, B. amyloliquefaciens, B. firmus* and *B. pumilus*)	*Pythium spp., Fusarium spp., Rhizoctonia solani, Aspergillus flavus*	Competition, direct antibiosis, induced resistance	Pertot *et al.*, 2015, Shafi J. *et al.*, 2017
2.	*Coniothyrium minitans*	*Sclerotinia sclerotiorum* and *S. trifoliorum*	Lysis by chitinase and β-1, 3 glucanase	Pertot *et al.*, 2015
3.	*Gliocladium catenulatum*	Species of *Rhizoctonia, Pythium, Phytophthora, Fusarium, Didymella, Botrytis, Verticillium, Alternaria, Cladosporium, Helminthosporium, Penicillium* and *Plicaria*	Toxin production	Pertot *et al.*, 2015
4.	*Purpureocillium lilacinum* QLP 12 (previously *Paecilomyces lilacinus*)	*Verticillium dahliae, R. solani*and nematodes	Parasitism	Lan *et al.*, 2017
5.	*Phlebiopsis gigantea*	*Heterobasidion annosum*	Competition for resources	Pertot *et al.*, 2015
6.	*Pseudomonas spp.*	*Pythium spp. R. solani*	Production of antibiotics, siderophores, volatiles	Pertot *et al.*, 2015
7.	*Pythium oligandrum*	Species of *Alternaria, Botrytis, Fusarium, Gaeumannonyces, Ophistoma, Phoma, Pseudocercosporella, Pythium, Sclerotinia* and *Sclerotium*	Hyperparasitism	Pertot *et al.*, 2015
8.	*Streptomyces* spp	Species of *Fusarium, Rhizoctonia, Phytophthora, Pythium, Phytomatotricum, Aphanomyces, Monosprascus, Armillaria, Sclerotinia, Verticillium, Geotrichum*	Mycoparasitism	Pertot *et al.*, 2015

Source: Panth, M. *et al.*, 2020.

10. Microbial Management

The application of biocontrol agents to the soil to suppress soilborne plant pathogens through parasitism, production of antagonistic chemicals, competition for the host for food and nutrients and the induction of resistance in plants against disease-causing pathogens (Shafique *et al.*, 2016). Several organisms are successfully used as biocontrol agents for controlling soilborne pathogens (Tables 10.3 and Table 10.4). The bacterial and fungal genera that are commercially employed as biological control agents include *Gliocladium, Bacillus, Coniothyrium, Paecilomyces, Phlebiopsis, Pseudomonas, Rhizobium, Serratia, Streptomyces,* and *Trichoderma* (Mazzola, *et al.*, 2017). The application of biocontrol agents in soil such as *T. viride, T. harzianum,* fluorescent *Pseudomonas* and *B. subtilis* have been found to be effective against root rot caused by soilborne plant pathogens in a number of crops (Shafique *et al.*, 2016; Loganathan, *et al.*, 2010). *Trichoderma* species are known to produce large quantities of fungi-toxic metabolites. They are the active mycoparasites which have been used as effective biocontrol agents against foliar and soilborne disease as well as plant parasitic soilborne nematodes (Shafique *et al.*, 2016; Kowsari, *et al.*, 2014).

Table 10.4: Management of Important Vegetable Diseases by different *Trichoderma* spp. (Singh, 2014)

Sl.No.	*Crop*	*Disease*	*Pathogen*	*Biocontrol Agent*
1.	Potato	Late Blight	*Phytophthora infestans*	*Trichoderma* spp
		Charcoal rot	*Macrophonina phaseolina*	*Trichoderma* spp.
		Black-scurf	*R. solani*	*T. viride*
2.	Tomato	Wilt	*Fusariun oxysporum* f. sp. *lycopersici*	*T. harzianum*
		Damping off	*Rhizoctonia solani, Pythiumaphanidermatum*	*T. harzianum*
		Root knot	*Meloidogyne incognita, M. javanica*	*T. harzianum*
3.	Chilli	Fruit rot and Die back	*Colletotrichum capsici*	*T. viride, T. harzianum, T. konningii, T. hamatum, T. longibrachiatum, T. pileatus*
		Root rot	*Sclerotium rolfsii*	*T. harzianum*
4.	Brinjal	Wilt	*Verticillium dahliae*	*T. viride, T. konningii*
		Damping off	*R. solani, P. aphani-dermatum*	*T. harzianum, T. hamatum, T. viride*
		Collar rot	*S. sclerotiorum*	*T. viride, T. virens*
5.	Cauliflower	Damping off	*R. solani, P. aphani-dermatum*	*T. viride, T. harzianum, T. konningii*
		Stalk rot	*S. sclerotiorum*	*T. harzianum*

Sl.No.	Crop	Disease	Pathogen	Biocontrol Agent
6.	Cucumber	Damping off and wilt	*R. solani, P. aphanidermatum, F. oxysporum fsp. cucumerinum*	*T. virens, T. harzianum*
		Root knot	*Meloidogyne incognita*	*T. harzianum*
7.	French bean	Root rot	*Rhizoctonia solani*	*T. virideT. hamatum*
8.	Okra	Wilt	*F. oxysporum vasinfectum*	*T. harzianum*
		Collar rot and root rot	*Sclerotium rolfsii*	*T. virideT. harzianum,*
9.	Bottle gourd	Collar rot	*Sclerotinia sclerotiorum*	*T. viride, T. virens*

The absence of an integrated approach to soil health and soil quality in general is the main cause of problems regarding soil-borne diseases. A lack of awareness and knowledge along the production chain, resulting in a lack of knowledge-based planning, monitoring and a lack of preventive measures leads to a reactive approach.

REFERENCES

Agrios N.G. (2005) Plant Pathology, 5th ed., Elsevier-Academic Press, pp. 635

Ara J., Sultana V., Qasim R., Ehteshamul-Haque S. and Ahmad V.U. (2005) Biological activity of *Spatoglossum asperum*: a brown alga. *Phytotherapy Research* 19 (7): 618–623.

Auger, J., Arnault, I., Diwo-Allain, S., Ravier, F., Molia, M. and Pettiti, M. (2004). Insecticidal and fungicidal potential of *Allium* products and substances as biofumigants. Agroindustria, 3(3): 367-370.

Baysal-Gurel F., Gardener B.M. and Miller S.A. (2012) Soilborne disease management in organic vegetable production. Available on: www.extension.org/pages/64951.

Blachinski D., Shtienberg D., Dinoor A., Kafkafi U., Sujkowski L.S., Zitter T.A. and Fry W.E. (1996) Influence of foliar application of nitrogen and potassium on Alternaria diseases in potato, tomato and cotton. *Phytoparasitica.* 24: 281–292.

Bonanomi, G., Antignani, V., Pane, C. and Scala, F. (2007) Suppression of soilborne fungal diseases with organic amendments. *J. Plant Pathol.* 89: 311–324.

Chase A.R. (1989) Effect of nitrogen and potassium fertilizer rates on severity of *Xanthomonas* blight of syngoniumpodophyllum. *Plant Dis.,* 73: 972–975.

Chitwood, D. J. (2003) Research on plant-parasitic nematode biology conducted by the United States Department of Agriculture-Agricultural Research Service. *Pest Manag Sci.*, 59 (6-7): 748-53.

Conn, K.L. and Lazarovits, G. (1999). Impact of animal manures on *Verticillium* wilt, potato scab, and soil microbial populations. *Canadian Journal of Plant Pathology*. 21(1): 81-92.

Dordas C. (2007) Role of nutrients in controlling plant diseases in sustainable agriculture. *A review. Agron. Sustain. Dev.*, 28 (2008) 33–46.

El-Sharouny, E.E. (2015). Effect of different soil amendments on the microbial count correlated with resistance of apple plants towards pathogenic *Rhizoctonia solani* AG-5. Biotechnology and Biotechnological Equipment, 29(3): 463-469.

Elad, Y., Katan, J. and Chet, I. (1980). Physical, biological, and chemical control integrated for soilborne diseases in potatoes. Phytopathology, 70(5), 418-422. doi: 10.1094/Phyto-70-418

Glaser, B., Lehmann, J., Zech, W. (2002) Ameliorating physical and chemical properties of highly weathered soils in the tropics with charcoal-a review. *Biol. Fertil. Soils*, 35: 219–230.

Graham D.R. and Webb M.J. (1991) Micronutrients and disease resistance and tolerance in plants, in: Mortvedt J.J., Cox F.R., Shuman L.M., Welch R.M. (Eds.), Micronutrients in Agriculture, 2nd ed., Soil Science Society of America, Inc. Madison, Wisconsin, USA, pp. 329–370.

https: //www.apsnet.org/edcenter/disimpactmngmnt/topc/Pages/ BiologicalControl.aspx

https: //www.apsnet.org/publications/phytopathology/backissues/ Documents/1973Abstracts/Phyto63_155.htm

Shafi, J., Tian, H. and Ji, M. (2017) *Bacillus* species are versatile weapons for plant pathogens: *A review. Biotechnol. Biotechnol. Equip.*, 31: 446–459.

Shafique, H.A., Sultana, V., Ehteshamul-Haque, S. and Athar, M. (2016) Management of soil-borne diseases of organic vegetables. *J. Plant Prot. Res.*, 56: 221–230.

Silber, A., Levkovitch, I. and Graber, E.R. (2010) pH-dependent mineral release and surface properties of corn strawn biochar: Agronomic implications. *Environ. Sci. Technol.*, 44: 9318–9323.

Singh, H.B. (2014) Management of plant pathogens with microorganisms. *Proc. Natl. Acad. Sci.*, 80: 443-454.

Gamliel, A., Austeraweil, M. and Kritzman, M. (2000). Nonchemical approach to soilborne pest management: Organic amendments. Crop Protection, 19(8): 847-853.

Hoffland E., Jegger M.J. and van Beusichem M.L. (2000) Effect of nitrogen supply rate on disease resistance in tomato depends on the pathogen, *Plant Soil* 218: 239–247.

Katan, J. (2000). Soil and substrate disinfestation as influenced by new technologies and constraints. *Acta Horticulturae*. 532: 29-38.

Katan, J., Greenberger, A., Alon, H. and Grinstein, A. (1976). Solar heating by polyethylene mulching for the control of diseases caused by soil-borne pathogens. Phytopathology, 66(5), 683-688. doi: 10.1094/Phyto-66-683

Keinath P.A. and Loria R. (1996) Management of Common Scab of Potato with Plant Nutrients, in: Engelhard W.A. (Ed.), Management of Diseases with Macro- and Microelements, APS Press, Minneapolis, USA, pp. 152–166.

Kostandi S.F. and Soliman M.F. (1991) Effect of nitrogen rates at different growth-stages on corn yield and common smut disease (Ustilage maydis (DC) Corda), *J. Agron, Crop Sci.*, 167: 53–60

Kowsari, M., Motallebi, M. and Zamani, R.M. (2014) Construction of new GFP-tagged fusants for *Trichoderma harzianum* with enhanced biocontrol activity. *J. Plant Prot. Res.*, 54: 122–131.

Krauss A. (1999) Balanced Nutrition and Biotic Stress, IFA Agricultural Conference on Managing Plant Nutrition, 29 June-2 July 1999, Barcelona, Spain.

Lagunas-Solar, M.C., Macdonald, J.D. and Granett, J. (1993). Control of Pests and Pathogens in Agricultural Soils with Radio Frequency Power. A Proposal for the Defense Technology Conversion, Reinvestment, and Transition Assistance Program. (Submitted to: Advanced Research Projects Agency). Davis: University of California.

Lan, X., Zhang, J., Zong, Z., Ma, Q. and Wang, Y. (2017). Evaluation of the biocontrol potential of *Purpureocillium lilacinum* QLP12 against *Verticillium dahliae* in eggplant. *Biomed Res. Int.*, 2: 1–8.

Larkin, R.P. and Griffin, T.S. (2007). Control of soilborne potato diseases using *Brassica* green manures. *Crop Protection*, 26(7): 1067-1077.

Larkin, R.P. and Honeycutt, C.W. (2006). Effects of different 3-year cropping systems on soil microbial communities and rhizoctonia diseases of potato. *Phytopathology*, 96(1): 68-79. pmid: 18944206

Loganathan, M., Sible, G.V., Maruthasalam, S., Saravanakumar, D., Raguchander, T., Sivakumar, M. and Samiyappan, R. (2010) *Trichoderma* and chitin mixture based bioformulation for the management of head rot (*Sclerotinia sclerotiorum* (Lib.) de Bary)–root-knot (*Meloidogyne incognita* Kofoid and White, Chitwood) complex diseases of cabbage. Arch. Phytopathol. *Plant Prot.*, 43: 1011–1024.

Marschner H. (1995) Mineral Nutrition of Higher Plants, 2nd ed., Academic Press, London, p. 889.

Mazzola, M., Granatstein, D.M., Elfving, D.C. and Mullinix, K. (2001). Suppression of specific apple root pathogens by *Brassica napus* seed meal amendment regardless of glucosinolate content. Phytopathology, 91(7), 673-679. pmid: 18942997

Mazzola, M. and Freilich, S. (2017) Prospects for biological soilborne disease control: Application of indigenous versus synthetic microbiomes. Phytopathology, 107, 256–263.

Nelson W.R. and Van Staden J. (1985) 1-Aminocyclopropane-1-carboxylic acid in seaweed concentrate. Botanica Màrina 28 (9): 415–417.

Panth M, Hassler S. C. and Baysal-Gurel F. (2020). Methods for Management of Soilborne Diseases in Crop Production. Agriculture, 10, 16, doi: 10.3390/ agriculture10010016

Pertot, I., Alabouvette, C., Esteve, E.H. and Franca, S. (2015) Focus group in soil borne diseases: The use of microbial biocontrol agents against soilborne diseases. In EIP-AGRI Focus Group on Soil-Borne Diseases: Final Report, Claude, A., Clarkson, J., de Cara, F.M., de Caravalho Franca, S., Debode, J., Elorrieta, M., Furlan, L., Grand, A., Hinarejos Esteve, E., Kos, J., *et al.*, Eds., EIP-AGRI: Brussels, Belgium, , pp. 3–5.

Rodríguez-Kábana, R. (1986). Organic and inorganic nitrogen amendments to soil as nematode suppressants. *Journal of Nematology,* 18(2): 129-34

Shafique, H.A., Sultana, V., Ehteshamul-Haque, S. and Athar, M. (2016). Management of soil-borne diseases of organic vegetables. *J. Plant Prot. Res.*, 56: 221–230.

Stephens, C. T. and Stebbins, T. C. (1985). Control of dampingoff pathogens in soilless container media. *Plant Disease,* 69(6): 494-496.

Stevens, C., Khan, V.A., Rodriguez-Kabana, R., Ploper, L.D., Backman, P.A., Collins, D.J. and Igwegbe, E.C.K. (2003). Integration of soil solarization with chemical, biological and cultural control for the management of soilborne diseases of vegetables. *Plant and Soil,* 253(2): 493-506.

Sultana V., Baloch G.N., Ambreen, Ara J., Tariq M.R. and Ehteshamul-Haque S. (2011a) Comparative efficacy of a red alga *Solieria robusta,* chemical fertilizers and pesticides in managing the root diseases and growth of soybean. *Pakistan Journal of Botany* 43 (1): 1–6.

Tariq A., Ara J., Sultana V., Ehteshamul-Haque S. and Athar M. (2011) Antioxidant potential of seaweeds occurring at Karachi coast of Pakistan. *Journal of Applied Botany and Food Quality.* 84 (2): 207–212.

Thies, J.E., Rillig, M.C. and Graber, E.R. Biochar effects on the abundance, activity and diversity of the soil biota. In Biochar for Environmental Management: Science and Technology, 2nd ed., Lehmann, J., Joseph, S., Eds.,

Welke, S.E. (2005) The effect of compost extract on the yield of strawberries and the severity of Botrytis cinerea. J. Sustain. Agric., 25, 57–68.

Woltz S.S. and Engelhar A.W. (1973) *Fusarium* wilt of chrysanthemum effect of nitrogen source and lime on disease development. *Phytopathology.* 63: 155–157.

Wu Y., Jenkins T., Blunden G., Whapham C. and Hankins S.D. (1997) The role of betains in alkaline extracts of *Ascophyllum nodosum* in reduction of *Meloidogyne javanica* and *M. incognita* infestations of tomato plants. Fundamental and Applied Nematology 20 (2): 99–102.

Xiao, C.L., Subbarao, K.V., Schulbach, K.F. and Koike, S.T. (1998). Effects of crop rotation and irrigation on *Verticillium dahliae* microsclerotia in soil and wilt in cauliflower. Phytopathology, 88(10), 1046-1055. pmid: 18944816

Zhang X. and Schmidt R.E. (2000) Hormone-containing products' impact on antioxidant status of tall fescue and creeping bentgrass subjected to drought. *Crop Science*, 40 (5): 1344–1349.

Chapter 11

Diseases Management of Tomato Crop in Organic Farming

Dushyant Kumar*, Joni Kumar, Arun Kumar and Krishna Pal

Chaudhary Chhotu Ram (P.G.) College, Muzaffarnagar, Uttar Pradesh
**e-mail: ccrdkpathology1992@gmail.com*

ABSTRACT

Tomato (Lycopersicon esculentum) is a major food source of carotenoids, vitamins and some other phenolics compounds. Tomato based industries are the most globalized and advanced industries. The maximum production of tomatoes is obtained from temperate zones. The leading producers of tomatoes are China, USA and India. In India, the production of tomatoes was very low as compared to China and USA. This is because of the heavy damage to tomatoes plant by several pathogens. Among the pathogens, Fusarium, Pythium, Rhizoctonia and Verticillium are the major pathogens which affect the quality tomatoes and also reduced the production of crop. Management of these pathogens by synthetic chemicals leads the negative impact on crops and also on associated organisms. There is need of alternate method for the sustainable management of tomato diseases and production of tomato crop.

Keywords: *Tomato, Disease, Organic farming, Management.*

INTRODUCTION

Tomato (*Lycopersicon esculentum*) belongs to the family Solanaceae. It is a major food source of carotenoids with estimated 80 per cent of daily intake (Willcox *et al.*, 2003). Several studies conducted on tomatoes have shown it beneficial effects of many diseases including cancer (Giovannucci *et al.*, 2002).

Apart from this, tomatoes contain many other beneficial nutrients like vitamin C and some phenolics (Willcox *et al.*, 2003). Al-Wandawi *et al.* (1985) reported that tomato peel contains high levels of lycopene compared to the pulp and seeds. In addition, tomato skin and seeds were reported to contain essential amino acids and the tomato seeds had higher amounts of minerals (Fe, Mn, Zn and Cu). They also have monounsaturated fatty acids especially oleic acid and linolenic acid (Raffo *et al.*, 2002). The tomato is most globalized and advanced industries. The maximum production of tomatoes is obtained from temperate zones. China, USA, and India are the leading producers of tomatoes (www.fao.org). In year 2020, the volume of tomato production in India amounted to over 21 million metric tons. This was an increase from the previous fiscal year. Andhra Pradesh, Madhya Pradesh and Karnataka were the leading producers of tomato across the country (Statista.com).

The production and quality of tomatoes are effected by pathogens (Ramyabharathi *et al.*, 2012). Among the pathogens that affect the tomato crop, soil borne fungal pathogens including *Fusarium, Pythium, Rhizoctonia,* and *Verticillium* causing the root rot or damping-off and wilt affect the quality with yield reduction (Lucas *et al.*, 1997). In spite of the promising results shown by chemical treatments that phytotoxicity and chemical residues are the major problems leading to environmental pollution and also for human health hazards. Organic crop production is partially characterized by the absence of synthetic pesticides and fertilizers but practices that promote ecosystem health are even more important.

What Types of Symptoms Appear on Tomato Plant?

1. **Fungi** cause spots, lesions, blights, wilts, rots, fruiting bodies, mildews, molds, leaf spots, root rots and blotches. Fungi are typically spread by wind, rain, soil, mechanical means and infected plant material.
2. **Bacteria** cause spots, wilts, rots, blights, cankers, exudates, galls, yellowing, leaf spots, watery blotches, wilting. Bacteria are typically spread by rain, mechanical means, planting material, vectors
3. **Viruses** cause mottling, leaf and stem distortions, mosaic patterns, rings and stunting. Viruses cause interesting symptoms, some are beautiful. Viruses are spread by mechanical means, vectors and in plant material.
4. **Nematodes** cause wilting, stunting, yellowing of entire plants. This is because the roots of the plant are infected and the plant is starving or thirsty. Nematodes are spread by soil on equipment or workers boots or on infected plant material.

Material-based Management Practices in Tomato Crop

1. Cultural Practices

- ☆ Promote healthy soils and healthy plants. Healthy soil is the hallmark of organic agriculture. An unhealthy plant is very attractive to diseases.
- ☆ Soils rich in organic matter are shown to increase soil biodiversity and help to create and abundance of beneficial soil microorganisms. Using compost has been shown to increase the suppressiveness of the soil by encouraging beneficial microorganisms as well as inducing disease resistance in plants by simply having healthier plants.

2. Exclusion

- ☆ Disease-free seeds, transplants or plant stock.
- ☆ Prevent introduction of diseased plants or soil.
- ☆ Disease free water source.
- ☆ Control insects that can carry disease.
- ☆ Soil solarization.
- ☆ Disease resistant varieties.
- ☆ Good sanitation from the prior season. Remove diseased plants or weeds from the field. Don't put disease plants or weeds in the compost pile. Many diseases are resistant to high heats or can become resistant. Some pathogens form resistant structures that can tolerate unfavorable conditions.
- ☆ Always work infested fields last and clean off equipment.

3. Crop Rotation

- ☆ Plants adapted to disease free area.
- ☆ Plant at proper depth (below crown or graft)
- ☆ Use only thoroughly composted material
- ☆ Improve air circulation by staking, trellis or pruning
- ☆ Water in the morning
- ☆ Avoid overhead irrigation if possible

4. Create an Unfavorable Environment for Pathogens

- ☆ Increase soil drainage
- ☆ Avoid low-lying areas
- ☆ Irrigation management- Drip Irrigation
- ☆ Mulches - Plastic or plant-based -Reduce splash dispersal of pathogens

5. Variety Selection

- ☆ Pick varieties that are appropriate for your area.
- ☆ Keep records of a cultivar's performance and the disease pressure each season.
- ☆ Local heirlooms are generally better suited for a particular region.
- ☆ Use tissue culture plants (small fruits, some cut flowers, perennials) if available. These plants are often disease indexed. This is especially important for viruses. This is a specialized area and there are not always tissue culture plants available.

6. Physical/Mechanical Controls

Physical and Mechanical controls are very important in insect management on organic farms. They are also very important for disease control, especially in perennial cropping systems like fruit trees, small fruits and tree nurseries.

- ☆ **Soil solarization**: this is a nonchemical technique which help in control many soil borne pathogens and pests. This simple technique captures radiant heat energy from the sun, thereby causing physical, chemical and biological changes in the soil. Transparent polyethylene plastic placed on moist soil during the hot summer months increases soil temperatures to levels lethal to many soil borne plant pathogens, weed seeds and seedlings (including parasitic seed plants), nematodes and some soil residing mites. Soil solarization also improves plant nutrition by increasing the availability of nitrogen and other essential nutrients.
- ☆ **Hot Water Seed Treatment**: The hot water seed treatment can help to decrease disease in seeds. Times and temps of seed treatment is 122 F for 25 min

Important diseases of tomato and their organic management are list below:

Table 11.1: List of Tomato Diseases with Management Practices

Sl.No.	*Disease Name*	*Causal Organism*	*Management Practices (Organic Crops)*
1.	*Phytophthora* root rot	*Phytophthora parasitica*	*P. lycopersici* suppressed in organic soil in bioassays, associated with high microbial activity, low N content, and high *actinomycete* diversity
2.	Damping-off	*Pythium aphanidermatum*	Bioassays with soil from organic and conventional field plots
3.	Root-knot	*Meloidogyne incognita*	Replicated long-term farming systems experiment; organic and low-input plots with winter cover crops, composted manure, and wider rotations in organic plots

Sl.No.	Disease Name	Causal Organism	Management Practices (Organic Crops)
4.	Fusarium wilt	*Fusarium oxysporum*	1. Lower leaves turn yellow and die. This spreads to other young leaves and entire plant wilts within a week. 2. Remove and burry crop debris. 3. Seed treatment-2gm *Trichoderma*/100gm seed. Seedling treatment - *Trichoderma* 5-10 gm/lit water-solution. 4. Apply 80 kg Neem cake/ac 5. Apply 200 kg compost treated with 2 kg of *Pseudomonas* or *Trichoderma*/ac.
5.	Bacterial wilt	*Ralstonia solanacearum*	1. Rapid and complete wilting of normal grown up plants. Lower leaves may drop before wilting. (Cut infected plant- immerse in water- a white streak of bacterial ooze comes out from cut ends) 2. Removal and burial of crop debris. 3. Apply 2 kg pseudomonas with 200 kg compost/ac. 4. Spray cow urine 5 lit + vermiwash 5 lit in 100 lit water/ac or spray Aloe or vitex extract.
6.	Southern blight	*Sclerotium rolfsii*	Plots with organic amendments and mulch
7.	Late blight	*Phytophthora infestans*	1. Remove infected plants, weeds and alternate host plants and burry. 2. Seed treatment-*Trichoderma*/Pseudo 4gm/100 gm seed 3. Soil application of 2kg 4. Pseudo/*Trichoderma* + 200 kg FYM/acre 5. Spray of Onion-garlic extract. 6. When heavy incidence sprays Bordeaux mix. 1 per cent
8.	Septoria leaf spot	*Septoria lycopersici*	Survey of organic and conventional farms; mixed cropping, residue management, and botanical pesticides at organic farms
9.	Powdery mildew	*Leveillula taurica*	Replicated experiment with conventional, low-input, and organic treatments in solarized soil in plastic greenhouses; humic acid as fertilizer, organicpesticides, sulfur, and copper sprays in organic plots
10.	Early blight	*Alternaria solani*	1. Small brownish leaf lesions on leaves and defoliation of older leaves. Remove infected leaves and burry deep. 2. Seed treatment-with *Trichoderma* or *Pseudomonas* 3 gm/100gm seed; Dip seedlings for 20 min Tricho/Pseudo solution (5gm/lit water) 3. Spray 5 per cent neem extract or 5 per cent garlic extract or Bourdeaux mix 1 per cent

Sl.No.	Disease Name	Causal Organism	Management Practices (Organic Crops)
11	Tomato spotted wilt	*Tomato spotted wilt virus*	Intensive fertilizer and insecticide treatments in conventional plots; organic treatments, plant extracts, copper fungicide in organic plots
12.	Tomatoringspot	*TomatoRing spot virus*	Replicated experiment with conventional, low-input, and organic treatments in solarized soil in plastic greenhouses; humic acid as fertilizer, organic pesticides, sulfur, and copper sprays in organic plots
13	Blossom end rot		1. It is a symptom of calcium deficiency in the soil and therefore in the plant. 2. In acidic soils, apply lime as recommended or bone meal @250gm/plant to increase pH levels and to enhance Calcium mobility. 3. Remove affected fruits to avoid secondary infections
14.	Tomato bushystunt	*Tomato bushystunt virus*	Replicated experiment with conventional, low-input, and organic treatments in solarized soil in plastic greenhouses; humic acid as fertilizer, organic pesticides, sulfur, and copper sprays in organic plots
15.	Tomato leaf curl disease	*Tomato leaf curl virus*	1. Plants are severely stunted. No flowering and fruiting at later stage. 2. Remove infected plants and burry. 3. Spray Neem seed extract 5 per cent or neem oil 3 ml/lit water. 4. Spray cow urine-buttermilk spray. (15 days old cow urine 500 ml and fermented butter milk 500 ml in 15 lit water) Spray this mixture-100 lit/ac

CONCLUSION

In this chapter, we summarized the work done on organic farming. The organic production of tomatoes can be obtained with the use of none-fungicidal methods such as culture practices, physical practices, use of biological agents and plant products. These methods are safe to non-target organisms and associated environment. These methods also support sustainable development of production. There is a need of awareness of these methods among farmer community. Further research is required to evaluate the impact of these methods on crop production in comparison with synthetic pesticides.

REFERENCES

A. Raffo, C. Leonardi, V. Fogliano, P. Ambrosino, M. Salucci and L. Gennaro, (2002). Nutritional Value of Cherry tomatoes (*Lycopersicon esculentum* Cv

Naomi F1) harvested at different ripening stages *Journal of Agricultural and Food Chemistry*, 50, pp. 6550-6556

Aparna, K., Pasha, M. A., Rao, D. L. N. and Krishnaraj, P. U. (2014). Organic amendments as ecosystem engineers: microbial, biochemical and genomic evidence of soil health improvement in a tropical arid zone field site. *Ecol. Eng.* 71: 268–77

Bulluck, L. R. and Ristaino, J. B. (2002). Effect of synthetic and organic soil fertility amendments on southern blight, soil microbial communities, and yield of processing tomatoes. *Phytopathology* 92: 181–89

Darby H.M, Stone A.G and Dick R. P. (2006). Compost and manure mediated impacts on soilborne pathogens and soil quality. *Soil Sci. Soc. Am. J.*, 70: 347–58

Giovannucci, E., Rimm, E. B., Liu, Y., Stampfer, M. J. and Willett, W. C. (2002). A prospective study of tomato products, lycopene, and prostate cancer risk. *Journal of National Cancer Institute* 94, 391–398. https: //www.statista.com/statistics/1039712/india-production-volume-of-tomatoes/

Klein, E., Katan, J. and Gamliel, A. (2011). Soil suppressiveness to *Fusarium* disease following organic amendments and solarization. *Plant Dis.* 95: 1116–23

Liu B, TuC, Hu S and GumpertzM, BeagleRistaino J. (2007). Effects of organic, sustainable, and conventional management strategies in grower fields on soil physical, chemical, and biological factors and the incidence of Southern blight. *Appl. Soil Ecol.*, 37: 202–14

Lucas, G. B., Campbell, C. L. and Lucas, L. T. (1997). Introduction to Plant Disease Identification and Management. New Delhi: CBS Pub. and Distributors. 364.

Miguel, J. Costa and Ep Heuvelink. (2007). Today's worldwide tomato. International Suppliers Guide

Ramyabharathi, S. A., Meena, B. and Raguchander, T. (2012). Induction of chitinase and b-1, 3- glucanase PR proteins in tomato through liquid formulated Bacillus subtilis EPCO 16 against Fusarium wilt. J Today's *Biol Sci Res Rev JTBSRR* 1(1): 50–60

Van Bruggen, A. H. C. and Termorshuizen, A. J. (2003). Integrated approaches to root disease management in organic farming systems. *Australas. Plant Pathol.* 32: 141–56

Willcox, J. K., Catignani, G. L. and Lazarus, S. (2003). Tomatoes and cardiovascular heath. *Critical Review in Food Science and Nutrition*. 43: 1–18.

Chapter 12

Insect Pests of Cole Crops and their Management

Joni Kumar[1], Yogandar Sain[2], Dushyant Kumar[3], Sunil Kumar Singh[4] and Arun Kumar[5]*

[1]Department of Ag. Entomology, [3]Department of Plant Pathology, [4]Department of Ag. Botany,
C.C.R. (P.G.) College, Muzaffarnagar–251001, Uttar Pradesh
[2]Department of Ag. Entomology, J.V.C, Baraut, Baghpat–250001, Uttar Pradesh
**e-mail: jonikumar68@gmail.com*

ABSTRACT

Cole crops are widely distributed groups of vegetable. In India Cauliflower and cabbage are cultivated on large area and considered as a major cole crop. There are several factors which decrease the yield of Cole crops in India. A total 31 insect pest has been reported from cabbage and cauliflower. Among them diamond back moth (Plutellaxy lostella), cabbage borer (Hellula undalis), cabbage butterfly (Pieris brassicae), painted bug (Bagrada cruiceferum), aphids (Myzus persicae, Lipaphis erysimi and Bravicoryne brassicae) and mustard sawfly (Athelia proxima) are the major one. The insect pests of cole crops are mostly managed with the help of insecticides. But continue exposer of pest to the same insecticides may lead the development of insecticide resistance. Identification and development of other alternatives methods is required at current time. Green pesticides are considered to be the best alternatives of insecticides as it is harmless to biodiversity, no risk of resistance development, and increased the development of beneficial insects. The use of Cotesia plutellae, Telenomus remus, Campoletis chlorideae and Cotesia glomeratus are found effective against P. xylostella, S. litura, H. armigera and P. brassicae, respectively. The application of Nuclear Polyhedrosis Virus (NPV) at rate of 250-500 L.E (Larval Equivalent)/ha found effective against P. xylostella, S. litura and H. armigera. In this paper we have compiled the current data on non- insecticidal management of insect pests of cole crops.

Keywords: *Cole crops, Insect pests, IPM, Biological control and Microbial control etc.*

INTRODUCTION

The term 'Cole crops' is originated from Latin word caulis meaning stem or stalk of plant. Cole crops includes cabbage (*Brassica olerace* var *capitata*), Cauliflower (*Brassica oleracea* var *Botrytis*), Brussels sprouts (*Brassica oleracea* var *gemmifera*), Broccoli (*Brassica oleracea* var *italica*), Kohlrabi (*Brassica oleracea* var *gongylodes*) and Kale (*Brassica oleracea* var *acephala*) (Kahn *et al.*, 2009). These crops are the members of cruciferae family which is well adopted the cool climate. The young plant of these crops specially cabbage can tolerant 0°C temperature. They can be grown on a wide range of soils however sandy loam soil is best if crop is planted early. Cole crops perform best between 6-8 PH. Cole crops are the good source of vitamin A and C. It also contains potassium and other minerals (Kahn *et al.*, 2009).

Cole crops are widely distributed groups of vegetable. The world top producers of these vegetables are China, India, United States of America, Spain, Mexico, Italy, UK, and Germany. In India Cauliflower and cabbage are cultivated on large area and considered as a major crop. The major states where cauliflower produced largely are West Bengal, Bihar, Uttar Pradesh, Odisha, Haryana, Madhaya Pradesh, Gujarat an Punjab (Dhaliwal, 2007). Among the states, West Bengal is leading producer of cabbage with the production of 2288.50 tonnes. With production of 1058.78 tonnes cabbage, Odisha is the second largest producing state of India. The productivity of cabbage and cauliflower in other states is very low as compared to West Bengal and Odisha (NHM, 2018).

There are several factors which decrease the yield of Cole crops in India. The availability of quality seeds, fertilisers and micronutrients, lack of knowledge of recommended practices could be the major reason of low productivity of crops. Lack of money to purchase the required materials is another reason of low productivity (Rathod *et al.*, 2020). Apart from this, constant attack of pathogens and insect pests on crops is also the major constraint of low productivity. A total 31 insect pest has been reported from cabbage and cauliflower. Among them diamond back moth (*Plutella xylostella*), cabbage borer (*Hellula undalis*), cabbage butterfly (*Pieris brassicae*), painted bug (*Bagrada cruiceferum*), aphids (*Myzus persicae, Lipaphis erysimi* and *Bravicoryne brassicae*) and mustard sawfly (*Athelia proxima*) are the major one (Lal, 1975).

The insect pests of cole crops are mostly managed with the help of insecticides. But continue exposure of same insecticides may lead the development of insecticide resistance (Brattsten, 1986). Such resistance against insecticides has been reported from *P. xylostella* (Attique *et al.*, 2006), *S. litura* (Ahmad *et al.*2008), *H. armigera* (Qayyum *et al.*, 2015), and *M. persicae* (Devonshire *et al.*, 1998). For the management of these resistance pests farmers have to apply more doses of insecticides. The excess use of insecticides may have negative impact on environment and biodiversity. Apart from this it also affects the biology and development of associated organisms such as parasites and predators (Kumar *et al.*, 2020). Other harmful effects of insecticides reported previously are phytotoxicity to plant (Braganca *et al.*, 2018), problems of insecticide residue in vegetables (González-Rodríguez *et al.*, 2008), contamination of drinking water (Anju *et al.*, 2010) and carcinogenic to human beings (Doe *et al.*, 1994).

Identification and development of other alternatives methods is required at current time. Green pesticides are considered to be the best alternatives of insecticides (Hill, 2014 and Lamine, 2011) as it is harmless to biodiversity, no risk of resistance development, and increased the development of beneficial insects such as parasites, predators and pollinators. It also reduced soil erosion and improves soil health. So many studies have been conducted on organic farming with the use of non insecticidal methods (Rathore, 2017, Nile *et al.*, 2019, and Lamine, 2011) including plant volatiles (Kumar and Paul, 2021, Kumar *et al.*, 2018, Kumar *et al.*, 2017a and Kumar *et al.*, 2017b).The major non insecticidal techniques currently used by the farmers are cultural practices, physical practices, mechanical practices, use of plant products, parasites, predators and microbial agents. Several studies formulated in past have shown the potential of parasitoids and predators against pests (Greathead and Greathead, 1992). A number of parasitoidshas been reported from insect insect pests of cole crops. Among them *Cotesia plutellae, Telonomus remus, Compoletis chloriedae* and *Cotesia glomeratus* are found effective against *P. xylostella, S. litura, H. armigera* and *P. brassicae*, respectively (Bhat, 2018). Kumar and Paul (2017) have recorded 80-85 per cent parasitisation aphids with parasitoid, *Dieretiellarapae*.

The application of Nuclear polyhedrosis virus (NPV) at rate of 250-500 L.E (Larval Equivalent)/ha found effective against *P. xylostella, S. litura* and *H. armigera* (Ansari *et al.*, 2012). *Bacillus thuringiensis* also known as B.t is a bacterial product which effectively control lepidopteran insects including *P. Xylostella, S. litura* and *H. armigera* (Bravo *et al.*, 2011). Some b.t toxin also found effective against aphids (Chougule *et al.*, 2013). Some other studies conducted with fungal products are found to manage the populations of Lepidopteron (Akutse *et al.*, 2019) and sucking pest (Kim *et al.*, 2011). In this paper we have compiled the current data on non insecticide management of insect pests of cole crops.

Major Insect Pests of Cole Crops

Diamond Back Moth (*Plutella xylostella*)

Diamond back moth (DBM) is a small moth with diamond shape marking on it back. It can be seen on every part of world and has reported in Asia, Europe, Australia, America and New Zealand. It is one of the most distributed of pest Lepidoptera order. It is polyphagous pest which feed on several vegetables crop belonging to different families. Although it is very small in size but it is a serious and most destructive pest of crucifer crops. The female of diamondback moth lay eggs on leaves and the yellowish eggs are oval in shape. The female mostly prefer damaged part of plant for egg laying. This might be due to attraction of female towards volatile compound released from damaged plant. The larvae scrape the green materials from plant in early developmental stage and then start mining the leaves. Skletonization of leaves can be noticed in the later stage of development of larvae. The life of diamondback moth is completed within 13-16 days (Chandramohan, 1994).

Cabbage Borer (*Hellula undalis*)

Hellula undalis is a major pest of crucifer vegetables belongs to pyralidae family. The size of the moth is very small measuring 6-9 mm. The female of this pest lay eggs singly or group on host plant. After hatching, the newly emerged larvae start mining of leaves. In later stage the larvae start migrating from one plant other and the damage the plant stem, leaves and fruit (Gaikwad *et al.*, 2018). The larvae attack on the all cole crops and their damage make them unmarketable. It widely distributed in all continent. In India it has been reported from all cabbage and cauliflower growing regions. The maximum damage by *H. undalis* has been reported from February to March month.

Cabbage Leaf Weber (*Crocidolomia binotalis*)

Among the insect pest reported from cole crops, *C. binotalis* is a major one. It has been reported from Africa, Asia, and Australia. The female generally lay eggs in overlapping manner. The newly hatched larvae feed on lower sides of leaves. After that move to other plant where they webbed the leaves with the help of silken thread. Multiple heads may be produced due the damage of meristems of plant by larvae (Bhat, 2018).

Cabbage Butterfly (*Pieris brassicae*)

P. brassicae belongs to Pieridae family of order Lepidoptera. It also well known by its another name *i.e.* large white butterfly (Kumar and Paul, 2018). The adult of butterfly is very attractive and generally whitish in colour. In India, it has emerged as a major pest of cabbage (Kumar *et al.*, 2018). The female butterfly laid eggs in cluster on leaves. After hatching, the larvae start feeding on crop gregariously and they dispersed on plant. Complete defoliation of leaves can be noticed in the later stage of larvae.

Tobacco Caterpillar (*Spodoptera litura*)

Spodoptera litura is most damaging and serious pest of tobacco but can feed on Cole crops. Apart from these crops it has been reported from 115 crops of different family. The female of *S. litura* laid eggs in groups and covered them with yellowish silken hairs. These eggs are known as egg masses. Female generally laid on leaf surface but it can also laid on surface of jar or rearing cloth. The newly hatched larvae can be seen in group. In the initial stage they scrape the green materials from leaf surface and defoliation of leaf can be noticed in later stage (Sundar *et al.*, 2018).

Painted Bug (*Bagrada cruciferum*)

B. cruciferum is important pest of Cole crop of pentatomidae family. It is found active throughout the year and cause serious damage to crop during winter season. The nymph and adult of these suck the sap from leaf and developing pods of plant. White spot at feeding site can be seen of the leaves of plant. The damage may even be more serious if plant attacked at seeding stage. Around 26 to 70 per cent infestation has been reported at seedling stage (Singh and Malik, 1993).

Aphid (*Brevicoryne brassicae, Lipaphis erysimi* and *Myzus persicae*)

In India the vegetables of Cole crops are attacked by three species of aphids namely green peach aphids, *Myzus persicae*, cabbage aphids, *Bravicoryne brassicae* and turnip aphids, *Lipaphis erysimi* (Kumar and Paul, 2018a). All three species has been reported from different states of India. The appearance of adults of *B. Brassicae* is slightly bigger than *M. Persicae*and *L. erysimi*.However the size and the appearance of *M. Persicae* and *L. erysimi* are same in early stages of development. It is very difficult to distinguish one from another. A considerable colour variation in may even be seen in *M. persicae*. The adults of *B. brassicae* are covered by waxy secretion.

Both nymph and adults of all three species are found to damaged the plant by sucking sap. The plant may turn yellow when they damaged the crop severely. These insects secrete sugar rich sticky material which is known as 'honey dew'. The secreted honey dew may cover the plant surface which effect photosynthesis process. Apart from this, all three species of aphids are known to transmit viral diseases in plant.

Integrated Pest Management Strategy for Cole Crops Pest

IPM is a system that in the context of the associated environment and the population dynamics of the species utilizes all the suitable techniques and methods in as compatible manner as possible and maintains the pest populations at levels below those causing economic injury (FAO, 1972).

Cultural Control

a. Deep summer Ploughing of field to expose the larvae and other hibernating stages to sunlight and birds

b. **Crop rotation:** Planting of non crucifer crops after the harvesting of main crop. Thie practice stop the food supply to larvae. This management practice is not useful against polyphagous pest such as *Spodoptera litura* and *Helicoverpa armigera.*

c. **Nutrients management:** Appropriate use of nitrogenous fertilisers because access use of these fertilisers may invite heavy incidence of sucking pests including aphids and painted bugs

d. **Variations in sowing times:** Early sowing of mustard (first week of November) May escape the incidence of aphids

e. **Intercropping:** Planting of other crop preferred by pest in row within the main crop. For instance mustard is planting within main crop cabbage to distract the insect from main crops.

f. **Trap cropping:** In this practice, the most preferred crops are planted around the main crop to attract target pest egg laying and feeding. *e.g.* planting of castor and marigolds around cabbage may attract *S. litura* and *H. armigera*, respectively. The female of *H. armigera* prefer marigold for eggs laying. The development of newly hatched larvae of *H. armigera* is effected by the chemical secretion of marigold.

Physical Control

a. **Use of light traps:** It is well known that lepidopteran insects are nocturnal in habits. They always get attracted towards the sources of light. Installation of 4-5 light traps/hectares are sufficient to captured the adult moths of *Plutella xylostella, S. litura, H. armigera, C. binotalis* and *P. brassicae*. Both male and females of thses species can be trapped in a jar which is already filled with kerosene or insecticides solution. This practice could be used with other methods of IPM.

b. **Yellow sticky traps:** This management practice is mostly employed against sucking pests like aphids. The yellow colour board of paper or plastic is smeared with sticky materials such as oil or greesh. Such 20-30 traps are need to be fixed in one hectare filed.

Mechanical Control

a. Collection and destruction of egg masses of *S.litura*

b. Collection and destruction of newly hatched gregarious larvae of *S. litura* and *P. brassicae*

c. Collection and destruction of aphids infested twigs or leaves of plant to kill the early populations aphids

Behavioural Control

In this practice certain chemicals which modify the behaviour of insects are employed to attract or direct the insect in desired location. These behaviour manipulating chemicals are known as semiochemicals. Semiochemicals are further divided into two groups *e.g.* pheromones and allelochemicals. In IPM pheromones are used in following ways:

a. Monitoring of pest populations with few numbers of pheromone traps *e.g.* 6-8 traps/ha. The monitoring of pest populations gives the ideas of pest species and their abundance which is very helpful for deciding further action of management. This management practice can be used against *P. xylostella, S. litura, H. Armigera* and *P. brassicae.*

b. ***Mass trapping:*** This practice is similar to monitoring. The only difference is that here the number of traps are higher than monitoring. More number of traps (20-25)/hectare are used to mass capture the populations of insects. This technique is also employed against lepidopteran insects such *S. litura, H. armigera.*

c. **Attract and infect:** In this management practice insects are attracted in pre planed locations with the help of pheromones. Where some viral (NPV, GV) or fungal (*Beauveriabassiana*) agents are already mixed with attractive chemicals. The insects captured in this pre installed traps get infected with viruses and the infected insects spread the virus to other populations while mating and feeding.

Use of Plant Products

Plant with insecticidal values are the best alternatives methods of conventional insecticides.

Biological Control

Use of living organisms for the management of insect pests is known as biological control. This management practice is further divided into two categories:

a. Use of Parasitoids and Predators

Different parasitoids and predators recommended against pest of cole crops are listed here:

Sl.No.	*Host insect*	*Recommended Biological Control Agents*	*References*
1	Diamond Back Moth, *Plutella xylostella*	*Chrysoperla carnea*	Reddy *et al.*, 2004
		Anomma nigricans	Goudegnon *et al.*, 2004
		Trichogramma brassicae	Babendreier *et al.*, 2003a
		Microplitis plutellae	Lim, 1986
		Cotesia pluetalla and Diadegma semiclausum	Ayalew *et al.*, 2004
		Tetrastichus sokolowskii	Liu *et al.*, 2000
		Apanteles piceotrichosus	Guilloux *et al.*, 2004
		Oomyzus sokolowskii	Kirk *et al.*, 2001
2	Cabbage borer, *Hellula undalis*	*Bassus* sp. *Trathala flavoorbitalis, Chelonus* sp. and *Phanerotoma sp*	Sivapragasam and Chua, 1997

Sl.No.	Host insect	Recommended Biological Control Agents	References
3	Cabbage butterfly, *Pieris brassicae*	*Cotesia glomerata*	Razmi *et al.*, 2011
		Pteromalus puparum	
		Exorista segregata	
		Agrothereutes adustus	
		Blapsidotes vicinus	
		Chrysoperla carnea	Fakharany and Hendawy, 2014
4	Tobacco caterpillar, *Spodoptera litura*	*Chelonusbifoveolatus Szpligeti, Coccygidium luteum, Cotesiaicipe, Meteorideatestacea*, and *Bracon sp., Anatrichus erinaceus*	Koffi *et al.*, 2020
		Telenomus remus	Martinez *et al.*, 2012
5	*Myzus persicae, Lipaphis erysimi, Brevicoryne brassicae*	*Diaeretiella rapae*	Kumar and Paul, 2018b
		Aphidius colemani	Prado *et al.*, 2015
		Coccinella transversalis, Coccinella septempunctata and *Menochilus sexmaculatus*	Sharma *et al.*, 2021
6	Mustard sawfly, *Atheliaproxima*	*Coccinella transversalis, Coccinella septempunctata* and *Menochilus sexmaculatus*	Sharma *et al.*, 2021

b. Use of Microbial Agents

Different microbial agents recommended against pest of cole crops are listed here:

Sl.No.	Name of Insect	Microbial Agents	References
1	Diamondback moth, *Plutella xylostella*	Granulosis Virus (GV)	Yang *et al.*, 2012
		Bacillus thuringiensis subspecies kurstakia	Jurat-Fuentes and Jackson 2012
		B. thuringiensis sub-species aizawaia	Mashtoly *et al.*, 2011
2	Cabbage borer, *Hellula undalis*	*Bacillus thuringiensis* subspecies kurstakia	Jurat-Fuentes and Jackson 2012
		B. thuringiensis sub-species aizawaia	Mashtoly *et al.*, 2011
3	Cabbage butterfly, *Pieris brassicae*	*Bacillus thuringiensis* subspecies kurstakia	Jurat-Fuentes and Jackson 2012
		B. thuringiensis sub-species aizawaia	Mashtoly *et al.*, 2011

Sl.No.	Name of Insect	Microbial Agents	References
4	Tobacco caterpillar, *Spodoptera litura*	Si Nuclear Polyhedrosis virus (NPV)	Yang *et al.*, 2012
		Bacillus thuringiensis subspecies kurstakia	Jurat-Fuentes and Jackson 2012
		B. thuringiensis sub-species aizawaia	Mashtoly *et al.*, 2011
		Nomura earileyi	Thakre*et al.* (2011)
5	Gram pod borer *Helicoverpa armigera*	Ha Nuclear Polyhedrosis virus (NPV)	Yang *et al.*, 2012
		Bacillus thuringiensis subspecies kurstakia	Jurat-Fuentes and Jackson 2012
		B. thuringiensis sub-species aizawaia	Mashtoly *et al.*, 2011
		Nomura earileyi	Thakre *et al.*, 2011
6.	Aphids *Lipaphis erysimi, Myzus persicae, Bravicoryne brassicae*	*Conidiobolus thromboides*	Papierok and Eilenberg 2012
		Lecanicillium longisporum	Down 2009
		Metarhizium anisopliae	Jaronski and Jackson (2012)
		Verticilium lacanii	Askary *et al.*, 1998

CONCLUSIONS

It is widely accepted that the indiscriminate use of conventional insecticides have harmful impacts on environment and biological organisms including human beings. The residue of these chemicals can be seen in food grains, vegetables, fruits and animals products. Such contaminated food materials if eaten by human beings may leads the development of several carcinogenic diseases. To avoid this farmer communities have to adopt alternatives methods of conventional insecticides. Awareness among farmers regarding the benefits of non insecticidal methods is still required. Further studies need to be conducted on the potential benefits of these alternatives methods in contrast to conventional insecticides.

ACKNOWLEDGEMENTS

Authors express their sincere thanks to the Principal and Management committee of CCR (P.G.) College Muzaffarnagar for supporting and encouraging while writing this chapter.

REFERENCES

Ahmad, M., Sayyed, A. H., Saleem, M. A. and Ahmad, M. (2008). Evidence for field evolved resistance to newer insecticides in *Spodoptera litura* (Lepidoptera: Noctuidae) from Pakistan. *Crop Protection*, *27*(10), 1367-1372.

Akutse, K. S., Kimemia, J. W., Ekesi, S., Khamis, F. M., Ombura, O. L. and Subramanian, S. (2019). Ovicidal effects of entomopathogenic fungal isolates on the invasive fall armyworm *Spodoptera frugiperda* (Lepidoptera: Noctuidae). *Journal of Applied Entomology, 143*(6), 626-634.

Anju, A., Ravi S, P. and Bechan, S. (2010). Water pollution with special reference to pesticide contamination in India. *Journal of Water Resource and Protection,* DOI: 10.4236/jwarp.2010.25050.

Ansari, M. S., Ahmad, N. and Hasan, F. (2012).Potential of biopesticides in sustainable agriculture. In *Environmental protection strategies for sustainable development* (pp. 529-595).Springer, Dordrecht.

Askary, H., Carriere, Y., Belanger, R. R. and Brodeur, J. (1998). Pathogenicity of the fungus *Verticillium lecanii* to aphids and powdery mildew. *Biocontrol science and technology, 8*(1), 23-32.

Attique, M. N. R., Khaliq, A. and Sayyed, A. H. (2006). Could resistance to insecticides in *Plutella xylostella* (Lep., Plutellidae) be overcome by insecticide mixtures?. *Journal of Applied Entomology, 130*(2), 122-127.

Ayalew, G. and Ogol, C. K. P. O. 2006. Occurrence of the diamondback moth (*Plutella xylostella* L.) and its parasitoids in Ethiopia: influence of geographical region and agronomic trait. J. Appl. Ent., 130(6-7): 343-348.

Babendreier, D., Rostás, M., Höfte, M. C. J., Kuske, S. and Bigler, F. (2003). Effects of mass releases of *Trichogramma brassicae* on predatory insects in maize. *Entomologiaexperimentalisetapplicata, 108*(2), 115-124.

Bhat, D. M. (2018). First report of Crocidolomiabinotalis (Zeller) (Lepidoptera: Crambidae) as a serious pest of some brassica vegetable crops in Kashmir. *Journal of Entomological Research,* 42(2), 255-258.

Bhat, D. M. (2018).Incidence and diversity of lepidopterous insect pests and their parasitoids (natural enemies) on cole crops at danderkhah location in Srinagar District (J and K, India). *International Journal of Entomology Research,* 3(2), 107-113.

Bragança, I., Grosso, C., Rede, D., Sousa, S. R., Lemos, P. C., Domingues, V. F. and Delerue-Matos, C. (2018). Ecotoxicological effects of insecticides in plants assessed by germination and other phytotoxicity tools. In *Biotic and Abiotic Stress Tolerance in Plants* (pp. 47-76). Springer, Singapore.

Brattsten, L. B., Holyoke, C. W., Leeper, J. R. and Raffa, K. F. (1986). Insecticide resistance: challenge to pest management and basic research. *Science, 231*(4743), 1255-1260.

Bravo, A., Likitvivatanavong, S., Gill, S. S. and Soberón, M. (2011). Bacillus thuringiensis: a story of a successful bioinsecticide. *Insect biochemistry and molecular biology, 41*(7), 423-431.

Chandramohan, N. (1994). Seasonal Incidence of Diamondback Moth, Plutella xylostella L. and its ParasiToids in Nilgiris. *Journal of Biological control, 8*(2), 77-80.

Chougule, N. P., Li, H., Liu, S., Linz, L. B., Narva, K. E., Meade, T. and Bonning, B. C. (2013). Retargeting of the Bacillus thuringiensis toxin Cyt2Aa against hemipteran insect pests. *Proceedings of the National Academy of Sciences, 110*(21), 8465-8470.

Devonshire, A. L., Field, L. M., Foster, S. P., Moores, G. D., Williamson, M. S. and Blackman, R. L. (1998). The evolution of insecticide resistance in the peach–potato aphid, Myzuspersicae. *Philosophical Transactions of the Royal Society of London. Series B: Biological Sciences, 353*(1376), 1677-1684.

Dhaliwal, M. S. (2007). Cole crops: cabbage (No. BOOK).Kalyani Publishers.

Doe, J. E. and Paddle, G. M. (1994). The evaluation of carcinogenic risk to humans: occupational exposures in the spraying and application of insecticides. *Regulatory toxicology and pharmacology, 19*(3), 297-308.

Down, R. E., Cuthbertson, A. G., Mathers, J. J. and Walters, K. F. (2009). Dissemination of the entomopathogenic fungi, *Lecanicillium longisporum* and *L. muscarium*, by the predatory bug, Oriuslaevigatus, to provide concurrent control of Myzuspersicae, Frankliniellaoccidentalis and Bemisiatabaci. *Biological Control, 50*(2), 172-178.

El-Fakharany, S. K. M. and Hendawy, A. S. (2014). Field studies on cabbage white butterfly, *Pierisrapae* (Linnaeus) and its associated parasitoid and predatory species in Egypt. *Egyptian Journal of Biological Pest Control, 24*(2), 437-444.

Gaikwad, A. D., Bhede, B. V., Bokan, S. C. and Bhosle, B. B. (2018). Seasonal incidence of major insect pests, natural enemies on cauliflower and their correlation with weather parameters. *J EntomolZool Stud, 6*(5), 952-6.

González-Rodríguez, R. M., Rial-Otero, R., Cancho-Grande, B. and Simal-Gándara, J. (2008). Occurrence of fungicide and insecticide residues in trade samples of leafy vegetables. *Food Chemistry, 107*(3), 1342-1347.

Goudegnon, A. E., Kirk, A. A., Arvantitakis, L. and Bordat, D. (2004).Status of the diamondback moth and *Cotesia plutellae*, its main parasitoid, in the Cotonou and Porto-Novo periurban areas of Benin. *Improving Biocontrol of Plutella xylostella*, 173-178.

Greathead, D. and Greathead, A. (1992). Biological control of insect pests by insect parasitoids and predators: the BIOCAT. *Biocontrol, 13*(4), 61N-68N.

Guilloux, T., Monnerat, R., Castelo-Branco, M., Kirk, A. A. and Bordat, D. (2004). Population dynamics of *Plutella xylostella* (Lepidoptera: Yponomeutidae) and its parasitoids in the region of Brasilia. In *Improving biocontrol of Plutella xylostella. Proceedings of the International Symposium, 21–24 October 2002* (pp. 184-189). France: Montpellier.

Gutierrez-Martinez, A., Tolon-Becerra, A. and Lastra-Bravo, X. B. (2012). Biological control of spodopterafrugiperda eggs using Telenomus remus Nixon in maize-bean-squash polyculture. *American Journal of Agricultural and Biological Sciences*, *7*(3), 285-292.

Hajek, A. E., Papierok, B. E. R. N. A. R. D. and Eilenberg, J. Ø. R. G. E. N. (2012). Methods for study of the Entomophthorales. *Manual of techniques in invertebrate pathology*, 285-316.

Hill, S. B. (2014). Considerations for enabling the ecological redesign of organic and conventional agriculture: A social ecology and psychosocial perspective. In *Organic farming, prototype for sustainable agricultures* (pp. 401-422).Springer, Dordrecht.

Jurat-Fuentes, J. L. and Jackson, T. A. (2012).Bacterial entomopathogens. *Insect pathology*, 265-349.

Kahn, B. A., Rebek, E. J. and Damicone, J. P. (2009).Cole crop production (broccoli, cabbage, and cauliflower).

Kim, J. J., Lee, M. H., Yoon, C. S., Kim, H. S., Yoo, J. K. and Kim, K. C. (2001). Control of cotton aphid and greenhouse whitefly with a fungal pathogen. *Biologcal Control of Greenhouse Pests*, 8-15.

Kirk, A. A., Mercadier, G., Bordat, D., Delvare, G., Pichon, A., Arvanitakis, L. and Rincon, C. (2001). Variability in Plutella and its natural enemies: implications for biological control. In *The management of diamondback moth and other crucifer pests.Proceedings of the Fourth International Workshop* (Vol. 26, p. 29).

Koffi, D., Kyerematen, R., Eziah, V. Y., Agboka, K., Adom, M., Goergen, G. and Meagher, R. L. (2020). Natural enemies of the fall armyworm, Spodopterafrugiperda (JE Smith)(Lepidoptera: Noctuidae) in Ghana. *Florida Entomologist*, *103*(1), 85-90.

Kumar, J. and Paul, B. (2021).Electroantennogram responses of *Eariasvittella*(Fabricius) to volatiles of cotton plant.*J. Exp. Zool. India.* 24 (2): 1709-1715.

Kumar, J. and Paul, B. (2017). Population dynamics of aphid, Myzuspersicae (Sulzer)(Hemiptera: Aphididae) on different Brassica species. *Agricultural Science Digest-A Research Journal*, *37*(1), 64-67.

Kumar, J. and Paul, B. (2018a). Response of Pieris brassicae (Linnaeus) (Lepidoptera: Pieridae) to Brassica extracts.Journal of Entomology and Zoology Studies.6(3): 1226-1228.

Kumar, J. and Paul, B. (2018b). SEASONAL PARASITIZATION OF APHID BY *Dieretiellarapae* (M'INTOSH) IN *Brassica* ECOSYSTEM.

Kumar, J., Paul, B. and Nebapure, S. M. (2019). Electrophysiological response of Dysdercuscingulatus (Fabricius)(Hemiptera: Pyrrhocoreidae) to cotton plant extracts and other host related compounds. *Proceedings of the National Academy of Sciences, India Section B: Biological Sciences, 89*(3), 793-800.

Kumar, J., Paul, B., Nebapure, S. M. and Shivashankara, K. S. (2017b). Differential electroantennogram response of male and female of Dysdercuscingulatus to okra plant volatiles. *Journal of Entomology and Zoology Studies.* 5, 197-201.

Kumar, J., Paul, B., Nebapure, S. M. and Singh, S. (2017a). Comparative GC-MS analysis of two Brassica rapa L. varieties for identification of volatile compounds. *Chem. Sci. Rev. Lett, 6*, 884-889.

Kumar, J., Sain, Y., Gautam, M. P., Kumar, S. and Kumar, S. (2020). Toxic effects of insecticides on beneficial insects.Learning Media Publication.217-221pp.

Kumar, V., Kaur, S., Kumar, J. and Gupta, Y. (2018).Development of white butterfly, Pieris brassicae L. in cabbage ecosystem. *Journal of Entomology and Zoology Studies, 6*(4), 1270-1273.

Lal, O.P. (1975). A compendium of insect pest of vegetables in India.*Bull. Entomol. Res.* **16**, 31-56.

Lamine, C. (2011). Transition pathways towards a robust ecologization of agriculture and the need for system redesign. Cases from organic farming and IPM. *Journal of rural studies, 27*(2), 209-219.

Lim, G. S. (1986). Biological control of diamondback moth. In *Diamondback moth management: Proceedings of the first international workshop* (pp. 159-171). Asian Vegetable Research and Development Center, Shanhua, Taiwan.

Liu, S. S., Wang, X. G., Guo, S. J., He, J. H. and Shi, Z. H. (2000). Seasonal abundance of the parasitoid complex associated with the diamondback moth, *Plutella xylostella* (Lepidoptera: Plutellidae) in Hangzhou, China. *Bulletin of entomological research, 90*(3), 221-231.

Mashtoly, T. A., Abolmaaty, A., El-Said El-Zemaity, M., Hussien, M. I. and Alm, S. R. (2011). Enhanced toxicity of Bacillus thuringiensis subspecies kurstaki and aizawai to black cutworm larvae (Lepidoptera: Noctuidae) with Bacillus sp. NFD2 and Pseudomonas sp. FNFD1. *Journal of Economic Entomology, 104*(1), 41-46.

Nile, A. S., Kwon, Y. D. and Nile, S. H. (2019). Horticultural oils: possible alternatives to chemical pesticides and insecticides. *Environmental Science and Pollution Research, 26*(21), 21127-21139.

Prado, S. G., Jandricic, S. E. and Frank, S. D. (2015).Ecological interactions affecting the efficacy of Aphidiuscolemani in greenhouse crops. *Insects, 6*(2), 538-575.

Rathod, B. S., Narvariya, R. K. and Shrivastava, A. (2020) Constraints in the Production and Marketing of Cabbage in Khandwa District of Madhya Pradesh.Current Journal of Applied Science and Technology.10.9734/cjast/2020/v39i3531056

Rathore, H. S. (2017). Green pesticides for organic farming: occurrence and properties of essential oils for use in pest control. In *Green Pesticides Handbook* (pp. 3-26).CRC Press.

Razmi, M., Karimpour, Y., Safaralizadeh, M. and Safavi, S. (2011).Parasitoid complex of cabbage large white butterfly *Pieris brassicae* (L.)(Lepidoptera, Pieridae) in Urmia with new records from Iran. *Journal of Plant Protection Research.*

Sarma, D., Saikia, D. K., Devee, A. and Borkakati, R. N. (2021). Diversity of Insect Pests and Predators of Cabbage Ecosystem in Different Phonological Stages of Cabbage. *Int. J. Curr. Microbiol. App. Sci, 10*(07), 427-433.

Singh, H. and Malik, V. S. (1993).Biology of painted bug (Bagradacruciferarum). *Indian Journal of Agricultural Science, 63*, 672-672.

Sivapragasam, A. and Chua, T. H. (1997).Natural enemies for the cabbage webworm, Hellula undalis (Fabr.)(Lepidoptera: Pyralidae) in Malaysia. *Researches on population ecology, 39*(1), 3-10.

Sundar, B., Rashmi, V., Sumith, H. K. and Sandhya, S. (2018). Study the incidence and period of activity of Spodoptera litura on soybean. *Journal of Entomology and Zoology Studies, 6*(5), 331-333.

Thakre, M., Thakur, M., Malik, N. and Ganger, S. (2011). Mass scale cultivation of entomopathogenic fungus Nomuraearileyi using agricultural products and agro wastes. *Journal of Biopesticides*, 4(2), 176.

Yang, M. M., Li, M. L., an Zhang, Y., Wang, Y. Z., Qu, L. J., Wang, Q. H. and Ding, J. Y. (2012). Baculoviruses and insect pests control in China. *African Journal of Microbiology Research*, 6(2), 214-218.

Chapter 13

Organic Management Practices for Diseases of Papaya

Abhishek Kumar[1], *Priyanka Yadav*[1], *Ajay Kumar Mishra*[1], *Anupam Kumar*[1], *Amit Kumar Yadav*[1] *and Rohit Gangwar*[2]

[1]*Department of Plant Pathology*, [2]*Department of Floriculture, Sardar Vallabhbhai Patel University of Agriculture and Technology, Modipuram, Meerut, Uttar Pradesh*

ABSTRACT

Papaya (Carica papaya L.) is a large herbaceous tropical plant that grown for a source of nutrients such as provitamin- A, carotenoids, vitamin C, folate and dietary fibre. Papaya have been infected by many diseases such as anthracnose, ascochyta leaf spot, damping off papaya, papaya mosaic etc. which have been shown to be very important factors in reducing yield and marketability of papaya. The rationale of this chapter to describe the symptoms, etiology, epidemiology and management measures for the diseases of papaya. The nature and frequency of these diseases depend on ecological conditions and successful management depends on a thorough awareness of the pathogen, host plant, environment and their interaction. In general, disease management strategies involve different practices that include plant resistance and prophylactic and curative measures. The papaya diseases related have diverse etiology divided into those with infectious and non-infectious etiology that affect the fruit and those affecting the plant. The information reported has been extracted from the available scientific literatures.

Keywords: *Papaya, Diseases, Causal organism, Management.*

INTRODUCTION

The papaya (*Carica papaya* L.) is native to the tropics of the Americas, perhaps from southern Mexico and neighbouring Central America. The papaya is a large, tree-like plant, with a single stem growing from 5 to 10 m tall, with spirally arranged leaves confined to the top of the trunk. The fruit is ripe when it feels soft and its skin attains an-amber to orange hue. India is the leading papaya producer, with a 38.61 per cent share of the world production followed by Brazil (17.5 per cent) and Indonesia (6.89 per cent). Papaya fruit is a source of nutrients such as provitamin- A, carotenoids, vitamin C, folate and dietary fibre. Papaya skin, pulp and seeds also contain a variety of phytochemicals, including lycopene and polyphenols. The ripe fruit of the papaya is usually eaten raw without skin or seeds. Diseases are a significant limiting factor for papaya production. The nature and Frequency of these diseases depend on local conditions and effective management depends on a thorough knowledge of the pathogen, host plant, environment, and their interaction. The precise identity of the causal agent is of paramount importance, and disease management options must be economical. Assessment of disease incidence or severity, and fruit loss are the key factors in determining the economics of disease management. In general, disease management strategies involve different practices that include plant resistance and prophylactic and curative measures. The papaya diseases related have diverse etiology, divided into those with biotic (infectious) and abiotic (non-infectious) etiology that affect the fruit and those affecting the plant.

1. Anthracnose

Anthracnose is considered one of the principle postharvest diseases of papaya occurring in all production regions in the world having been noted to cause large losses in Brazil, Hawaii (USA), Mexico and India. The fungus infects various parts of the plant but its greatest importance occurs in fruits which become unfit for commercialization. The latent infection not detected at harvest, it develops after harvest principally during transport of the fruits for export or local markets. The absence of control measures and use of inappropriate postharvest procedures can result in production of up to 100 per cent diseased fruits from some orchards.

Etiology

The causal agent of anthracnose is *Colletotrichum gloeosporioides* (Penz.). It is a common in diverse tropical fruit plants have been reported in the teleomorphic form in fruits and leaves (Ram, 1984; Costa *et al.*, 2001 and Holliday, 1980). The fungus forms sub-cuticular and sub-epidermal acervuli, separate or confluent with pinkish to orange conidial masses that cover the lesion centers with setae. Conidiophores are septate, pale brown with hyaline conidia with obtuse ends or ellipsoidal with a rounded apex and a narrow, truncate base, aseptate, more or less guttulate to cylindrical. **Costa *et al.* (2001)** reported the occurrence of the perfect stage (*Glomerella cingulata*) of pathogen with perithecia on dead leaves and fruits of the host in a manner of solitary or aggregated, globose to obpyriform, dark brown to black in colour and asci with 8 spores, clavate to cylindrical.

Symptoms

The disease prominently appears on green immature fruits. The disease symptoms are in the form of brown to black depressed spots on the fruits. The initial symptoms are water-soaked, sunken spots on the fruit. The centres of these spots later turn black and then pink when the fungus produces spores. The flesh beneath the spots becomes soft and watery, which spreads to the entire fruit. Small, irregular-shaped water-soaked spots on leaves may also be seen. These spots eventually turn brown. On the fruits, the symptoms appear only upon ripening and may not be apparent at the time of harvest. Brown sunken spots develop on the fruit surface, which later on enlarge to form water soaked lesions. The flesh beneath the affected portion becomes soft and begins to rot.

Epidemiology

The infection generally occurs in the field during the initial stages of development of the fruits. The pathogen is remaining quiescent until the maturation phase of the fruit. Fruits and senescent leaves infected by *C. gloeosporioides*, principally the petioles, both on the plant as well as those that fall on the ground represent important sources of inoculum of the pathogen. The production of ascospores in the petioles of dry leaves has been frequently observed but the importance of the sexual phase in the cycle of the disease still is not clear. In the field, the production of conidia occurs in large quantity principally in the petioles of senescent leaves and in mature fruits, these being considered the source of inoculum. However, additional studies are required to understand their epidemiological importance. Drops of rain and splashes of irrigation water are the principle sources of dissemination of inoculum of the pathogen. When the conidia germinate (6-8 hours), they form appressoria (10-12 hours) on the surface of the fruits and by enzymatic action penetrate the cuticle. The hyphae remaining sub-cuticular in a latent state until the beginning of maturation of the fruits, growth resumes and symptoms of the disease appear. Conidia dispersed in the initial phases of development of the fruits have less chance of survival than those that are deposited on developed fruits and that remain and form the quiescent infections or infect the fruits after harvest. Climatic conditions that favour the incidence of anthracnose are temperatures near 28°C, ranging from 20 to 30°C, and relative humidity of the air more than to 95 per cent. The conidia require water in the free state to germinate and are liberated from the acervuli only when the relative humidity is greater than 95 per cent (Quimio, 1973).

Management

The management of anthracnose in the field should begin with the choice of the field site taking into consideration the prior history (old plantings), avoidance of excess humidity and conditions that favour the development of the disease, cultural practices, the reduction of inoculum and genetic resistance.

Measures adopted during the phases of production and postharvest processing of the fruits (care in handling, cleanliness of packaging and of the environment, control of storage temperatures and use of heat treatments (Ventura, 1995, Ventura and Costa, 2002).

Tatagiba *et al.* (1998a) reported that greater doses of boron and calcium (0.77g of B and 50g of Ca/plant) an increase of approximately 70 per cent in the incidence of anthracnose in fruits than the required amount. The incidence in fruits reached 100 per cent with a content of boron in the plant of 50mg/kg; the average content generally required being 23mg/kg. In another experiment conducted in the northern region of Espírito Santo by Incaper, where irrigation by microaspersion was utilized, a negative relation between the amounts of water utilized and the incidence of anthracnose was observed that demonstrating the possibility of irrigation management for control of the disease. The amount of 120mm reposition the evapo-transpiration besides contributing to the greatest production of fruits was where the least incidence of anthracnose was observed. In the check without application of water, the incidence of the disease in fruits reached 100 per cent (Tatagiba *et al.*, 2001).

Hydrothermal Treatment

Heat treatments have been used to control fungal diseases. A treatment by double immersion (30 minutes at 42°C, followed by 20 minutes at49°C) was used in the Hawaii when combined with regular application of fungicides in the field (Alvarez and Nishijima, 1987). In Hawaii the effect of disinfestation of fruits was studied under commercial conditions using forced air treatment of different temperatures followed by application of the fungicide prochloraz. Fruits treated with temperatures between 47.5°C and 48.5°C caused the least loss of firmness of the pulp. This effect appears to be associated with a decrease in the enzyme polygalacturonase, responsible for the catalyzation of the hydrolysis of pectins, components of the cell wall (Chan *et al.*, 1981; Lay-Yee, *et al.*, 1998).

2. Ascochyta Leaf Spot

Ascochyta leaf spot or leaf blight is caused by *Ascochyta caricae-papayae* fungal pathogen. It is responsible to cause approximately 30 per cent losses in yield as observed in some orchards of the northern region of Espírito Santo, principally during periods of high frequency of rain and when outbreaks of two-spotted spider mites (*Tetranychus urticae*), which cause lesions in the leaves that serve as ports of entry for the fungus occur.

Symptoms

The fungus preferentially infects younger leaves and fruits but also was reported to cause spots on flowers and young fruits. Penetration is highly favoured by the presence of injuries on the borders of leaves where the round lesions that present a dark-brown coloration and pycnidia arranged in concentric rings are generally observed. In the more advanced stage, when the conditions are highly favourable, the disease can cause a severe blight of the leaves in the apex of the stem leaving the plants unproductive and in some cases provoking the death of the apex. Under humid conditions, it is common to see light-colored tendrils of pycnidiospores oozing from pycnidia. Infections on flowers and young fruits are initially brown becoming dark and sunken. Initial infection of stem-end rot is through the broken peduncle, where early symptoms are a slight browning. The infected area is delineated by a narrow, light brown, translucent margin and as the infected tissue ages; it becomes wrinkled, dry and black. In the fruits, the lesions generally occur in the phase of maturation, near the peduncle, turning the tissue black, wrinkled and dry and frequently covered by a spongy mass of gray colour that tends to increase in size with the formation of pycnidia separate and embedded in the diseased tissue as the lesions age.

Epidemiology

Climatic conditions that favour the incidence of Ascochyta spot are temperatures between 21 and 26°C and high relative humidity (>90 per cent). On leaves, the disease is favoured by the occurrence of strong winds. In fruits, handling in harvest and postharvest causes injuries that enable the penetration of the pathogen. The fungus colonizes senescing leaves and petioles and produces abundant fruiting structures (pycnidia) in dead leaves that serve as a source of primary inoculum in the field. In older leaves, the presence of the burn of the leaves associated with the damage caused by tetranychid mites, principally the two-spotted spider mite (*Tetranychus urticae* Koch), owing to the wounds that they cause to the leaves, generally can be observed. Infection by *P. caricae-papayae* in mature fruits (more susceptible) requires a superficial wound to begin infection and a minimum of 6 hours of continual wetness (Sanchez *et al.*, 1991). In green fruits, the optimal temperature for infection is 21°C in the presence of water in the free state and on the mature fruits, the temperature range of 18 to 24°C enables the greatest development of lesions.

Management

Wounds created during harvest and postharvest handlings are quickly colonized during storage if conditions for infection exist. Since, the fruit are susceptible to chilling injury at temperatures below 7°C, normal storage and shipping temperatures are suitable for spore germination and infection. In Hawaii, Nishijima (1998b) reported that hot water treatment at 48°C for 20 min also reduced stem-end rot. Double hot water dip and vapor heat quarantine treatments also control the disease on fruits. Based on epidemiological characteristics of the disease, the use of sprinkle irrigation favours the sporulation of the fungus and its dissemination in the orchard. The removal of infected leaves, petioles and fruits is recommended for management of the disease. Equally, the prevention of injuries to the fruits during harvest is recommended. In an evaluation of genetic resistance, the germplasm of C. papaya and *C. cauliflora* was susceptible to the fungus (Sanchez *et al.*, 1991).

3. Damping Off Papaya (*Pythium aphanidermatum*)

Damping off occurs generally in sporadic form in plants in seedbeds or sacks in nurseries but may also occur in seedlings recently transplanted in the field. In the nurseries, papaya seedling roots are very susceptible during the first three months after seedling emergence.

Symptoms

Symptoms of the disease are initially in the form of a watery spot in the region in the collar of the plants that increases in size provoking the destruction of the tissues and, in consequence, lodging and death of the seedlings occurs. Symptoms were underdevelopment, yellowing of the leaves and subsequently, abscission of the older leaves with newer leaves remaining on the extremities of the plants and mummification of the fruits also occurring. In these plants, the root system presented a death of the lateral roots with a zone of reddish coloration occurring in the region of transition between the healthy and diseased parts (Ram *et al.*, 1983).

Epidemiology

The incidence of damping off is favoured by clayey, poorly drained soils with inadequate aeration, and by seeding densely and deeply. High temperatures and rainy periods contribute to an increase in the severity of the disease. The use of mulch also can significantly favour the incidence of the disease in young plants (Elder *et al.*, 2000).

Management

Various measures should be adopted to control damping off, beginning with the choice of the local of establishment of the nursery that should be in a well-ventilated location, free of flooding, good exposure to the sun, away from roads and fields of papaya (Marin and Gomes, 2000). Balanced fertilization with macro and micronutrients (N: P: K: Zn: B: S @ 100: 100: 100: 10: 5: 30 g/plant) reduced the level of root rot disease and completely controlled the bumpiness disorder in papaya fruit (Gupta *et al.*, 2016). Before sowing, seeds should be treated with fungal culture of *Trichoderma viride* (5 g/kg of seed) to protect the newly emerging seedlings. Soil amendments with solarization, application of neem cake + *Trichoderma harzianum* are effective for control of root rot disease (Reddy *et al.*, 2012).

4. Papaya Mosaic

Papaya ringspot virus is responsible to cause as papaya mosaic and papaya distortion ringspot that is one of the most important and destructive diseases of papaya, being one of the factors limiting development of its cultivation worldwide. In Brazil, the disease was recorded in 1935 (Bitancourt, 1935). The typical mosaic caused by potexvirus so far not found in India.

Symptoms

Early symptoms appear first on young leaves in the crowns of plants. The leaves exhibit a yellowing and later present an aspect of mosaic, green areas mixed with areas of yellow shade of variable form and size and well defined outline, resulting in a diminution in the rate of growth of the plants, loss of vigour and decreased fruit yield. Mottling and intense deformations and blisters may occur characterized by elevated areas of normal green color in contrast with remaining yellow areas. In plants infected by severe strains of the virus, the leaves are greatly distorted and may develop a finger like or shoestring appearance. In the region of the stem, irregular water soaked spots with an oily appearance may be present, these symptoms being greatly characteristic of the disease. Streaks on the petioles are usually lighter coloured and blend into normal color of the petiole. The canopy of the diseased plants becomes smaller with the progress of the disease, the leaves remaining being of reduced size and often deformed. The fruit surface may present indistinct small concentric rings, very distinct, with the centre green. The first ring spots are dark green to tan and about 1mm in diameter. In the later stages, several spots may join to form irregular shaped spots up to 8 mm in diameter and the rings may become necrotic and pale. Fruit yields in infected plants are markedly lower and fruits that mature after the tree is infected may appear normal but are of much low quality and generally unmarketable.

Spread and Causes

The virus is flexuous particle of 680-760 X12nm and virus particles consists of positive sense monopartite ssRNA. PRSV is a potyvirus belonging to Potyviridae. PRSV is transmitted by several aphid species of which the following are important *Aphis gossypii, A. craccivora, Rhopalosiphum maidis* and *Myzus persicae* all of which transmit the virus in a non-persistent manner. Papaya and Pumpkin are the major primary and secondary source of inoculums while rapid secondary spread can occur very fast leading to whole plantations getting totally infected. This would happen in plantations where young plants are in the vicinity of infected plants where populations of winged aphids are high. Transmission of PRSV-P is almost entirely due to transitory aphid populations, as papaya is not a preferred host for aphids and so colonies are very rarely found on plants.

Papaya ringspot virus presents a very rapid speed of dissemination from the focal point of the disease and all of the plants of an orchard may be infected after a period of 3-7 months as may be observed in areas with high winged aphid populations. The plants may be infected in any phase of vegetative development and the symptoms appear between 2-3 weeks after inoculation. However, research in India demonstrated a significant effect of the season of planting on the incidence of disease in the plants (Singh and Singh, 1998). Plants infected at a very young age never produce fruit but rarely die prematurely. There are reports that some isolates from Taiwan cause wilting and sometimes death of the plants (Nishijima *et al.*, 1989).

Management

Once established in a papaya orchard, the virus is rapidly disseminated and difficult to control as a consequence of its nonpersistent mode of transmission by aphids. Attempts to reduce disease levels by applying insecticides have not been successful. To control papaya ringspot virus the use of roguing and a combination of different methods is required (Costa *et al.*, 2000; Ventura and Costa, 2002). Considering that commercial varieties resistant to mosaic do not exist and control attempts by use of tolerant varieties and cross protection with attenuated strains of the virus have not shown success in the control of the disease. It has been possible to live economically with the virus using preventative methods and management of the culture aimed at reducing its dissemination.

Carry out periodic visits to the orchard and cut down all infected plants (rouging) as soon as symptoms of mosaic can be recognized. After plants dry or die they no longer will attract aphids. Following practices can be apply for effective management:

i. The trees should be cut as low as possible and do not allow re-growth from the stumps of infected trees. If necessary, spray the stumps with an herbicide (glyphosate) to prevent re-growth.
ii. Install nurseries and orchards as far as possible from other orchards, especially if mosaic occurs in that region.
iii. Eliminate the presence of cucurbits (such as pumpkin, melon, watermelon, cucumber, squash or others) which are hosts of the virus as well as hosts of the aphid vectors.
iv. Realize balanced fertilization and maintain the orchard clean to prevent the formation of aphid colonies in weeds.
v. Avoid planting rows in the same direction as predominant winds which can favour dissemination of aphids into, within the orchard and to nearby orchards.
vi. Destroy abandoned orchards, principally those with plants infected by the virus. One diseased tree can infect a whole neighbourhood.
vii. Buy papaya plants that have been grown under conditions that will prevent infection of seedlings or grow your plants from certified seeds.

CONCLUSION

Diseases of papaya have remained an economically significant in India and worldwide. Farmers lose millions of dollars annually due to several diseases of papaya. Rigorous seed certification and testing programs in developed countries have limited the impact of some viral and bacterial diseases within their value chains while developing countries commonly lack these safeguards. Lack of certified disease-free planting material in many developing country contexts contributes to further distribution of these pathogens. For tackle of above problems can be achieved by training farmers in proper production practices, on-farm management tools, using healthy seed tubers and planting in clean soils. Additional factors to consider in controlling these diseases can include sanitation, cultural practices, crop rotation with non host plants and the use of tolerant or resistant varieties.

REFERENCES

Alvarez, A.M. and Nishijima, W.T. (1987). Postharvest diseases of papaya. *Plant Disease* 718: 681-686.

Bitancourt, A.A. (1935). Mosaico do mamoeiro. O Biologico, 1: 41.

Camarch, R.F.A. and Lima, J.A.A. (1997) sobrevivencia do "Papaya Lethal Yellowing Virus" *Fitopatologia brasileira,* 22: 332-333.

Chan, H. T., Tam, Y. T. and Seo, S. T. (1981). Papaya polygalacturonase and its role in thermally injured ripening fruit. *Journal Food Sci.* 46: 190-191.

Costa, H., Ventura, J. A. and Tatagiba, J. S. (2000). Mosaico do mamoeiro: Uma ameaca a cultura no Espirito Santo. Vitoria: Incaper, 4pp.

Costa, H., Ventura, J. A. and Tatagiba, J. S. (2002). Severidade da antracnose e podridao peduncular do mamao no estado do Epirito Santo. *Fitopatologia Brasileira.* 27: 98.

Costa, H., Ventura, J. A., Rodrigues, C. H. and Tatagiba, J. S. (2001). Ocorrencia e patogenicidade de Glomerella cingulata em mamao no Norte do Estado do Espirito Santo. *Fitopatologia Brasileira,* 26: 328-328.

Dickman, M.B. and Alvarez, A.M. (1983). Latent infection of papaya caused by *Colletotrichum gloeosporioides. Plant Disease.* 67: 748-750.

Elder, R.J., Macleod, W.N.B., Reid, D.J. and Gillespie, R.L. (2000). Growth and yield of 3 hybrid papaya (*Carica papaya* L.) under mulched and bare ground conditions. *Australian Journal of Experimental Agriculture.* 40: 747-754.

Estrada, A.B., Dodd, J.C. and Jeffries, P. (2000). Effect of humidity and temperature on conidial germination and appressorium development of two Philippine isolates of mango anthracnose pathogen *Colletotrichum gloeosporioides. Plant Pathology.* 49: 608-618.

Gupta, A. K. and Chaudhary, R. (2016). Study on incidence and management of papaya disorder and diseases through balanced nutrition and common practices under NEPZ of India. *Hortflora Res. Spect.* 4: 380-82.

Holliday, P. (1980). Fungus diseases of tropical crops. Cambridge: Cambridge University Press, 607.

Lay-Yee, M., Clare, G. K., Petry, R. J., Fullerton, R. A. and Gunson, A. (1998). Quality and disease incidence of 'Waimanalo Solo' papaya following forced heat treatments. *Hort. Science,* 33 (5): 878-880.

Marin, S. L. D. and Gomes, J. A. (2000). Tecnicas de Cultivo do Mamao. Fortaleza-CE: FRUTAL, 57.

Nishijima, K.A., Couey, H.M. and Alvarez, A.M. (1987). Internal yellowing, a bacterial disease of papaya fruits caused by Enterobacter cloacae. *Plant Disease.* 71: 1029-1034.

Nishijima, W.T. (1998a). Miscellaneous papaya diseases. In: "Compendium of tropical fruit diseases", (eds. Ploetz, R.C., Zentmeyer, G.A., Nishijima, W.T., Rohrbach, K.G., Ohr, H.D.) 2nd ed., St. Paul. *American Phytopathological Society.*69-70.

Nishijima, W.T. (1998b). Papaya diseases caused by fungi: dry rot. In: "Compendium of tropical fruit diseases", (eds. Ploetz, R.C., Zentmeyer, G.A., Nishijima, W.T., Rohrbach, K.G., Ohr, H.D.) 2nd ed., St. Paul: *American Phytopathological Society.* 59-60.

Quimio, T.H. (1976). Pathogenicity and cultural characteristics of *Fusarium solani* from papaya. Kalikasan Philippine *J. of Biol.* 5: 241-250.

Ram, A. (1984). Forma perfeita de Colletotrichum gloeosporioides em frutos e folhas de mamoeiro na Bahia. *Fitopatologia Brasileira,* 9: 319.

Ram, A., Oliveira, M.L. de and Sacramento, C.K. (1983). Podridão das raízes do mamoeiro associada com Pythium sp. na Bahia. *Fitopatologia Brasileira* 10 (2): 581.

Reddy *et al.* (2012). Disease Management and Diagnostic Solutions, Papaya solution for crop management.

Sanchez, M., Dianese, J.C. and Costa, C.L. (1991). Fatores determinantes do dano de Phomacaricae-papayaeao fruto do mamoeiro (Carica papaya) e detecção de resistência ao fungo em 266 Carica gaudotiana. *Fitopatologia Brasileira.* 16 (1): 121-129.

Singh, A.L. and Singh, G. (1998). Effect of time of planting on growth fruiting behaviour and sex relation of papaya (Carica papaya). *Indian Journal of Agricultural Science.* 68(12): 769-772.

Tatagiba, J. S., Silva, J. G. F., Costa, H. and Ventura, J. A. (2001). Influência da irrigação na incidência da antracnose em frutos de mamão. *Fitopatologia Brasileira.* 26: 329.

Ventura, J. A. and Costa, H. (2002). Manejo integrado das doenças de fruteiras tropicais: abacaxi, banana e mamão. In: "ZAMBOLIM, Laercio. (Org.). Manejo integrado de doenças e pragas: fruteiras tropicais. Viçosa-MG", pp.279-352.

Ventura, J. A. (1995). Controle de doenças em pós-colheita de frutos tropicais. *Fitopatologiabrasileira*. 20: 273

Chapter 14

Organic Production of Asparagus

Vikash Kumar Yadav and Sharvan Kumar

Assistant Professor, Department of Agriculture, Invertis University, Bareilly, Uttar Pradesh

ABSTRACT

Asparagus is the most important minor vegetable crops which is grown by the farmers in small areas. Organic asparagus manufacturing, like different vegetable production systems, involves developing a practical device that provides adequate fertility while retaining powerful weed management. A nicely managed asparagus stand may additionally stay effective for 15 years or greater. In less warm districts, the asparagus fern will 'die off 'at some stage in iciness in similar style to a deciduous tree. The fern is then mulched into the soil or removed in coaching for harvest of the brand new asparagus spears in spring. The fern is not removed until completely 'dead' so that the asparagus root or crown is replenished as nutrients are reabsorbed from the stem and foliage.

***Keywords**: Asparagus, Production, Organic, Vegetable.*

INTRODUCTION

Asparagus has few pest or disorder problems and can be grown without artificial insecticides, making it a noticeably clean crop to grow organically. Properly weed control, in particular during established order, is essential to sell healthy boom and exceptional yield and first-rate. Organic asparagus manufacturing, like different vegetable production systems, involves developing a practical device that provides adequate fertility while retaining powerful weed management. A nicely managed asparagus stand may additionally stay effective for 15 years or greater. An knowledge of the once a year increase habit will assist in making plans an organic control program. Growth addiction asparagus is a perennial plant. It produces dark green fern-like foliage all through summer and in cooler climates becomes dormant over winter months. Asparagus spears start as leaf buds below the soil floor, elongating to form the spears above ground as temperatures warm inside the spring. Those are then cut at some stage in harvest, making way for brand new spears to provoke from buds on the crown beneath the soil. Harvesting of the spears maintains until spear pleasant deteriorates (throughout warm climate) or market charges fall. When harvesting ceases, the spear is permitted to develop into its completely improved leaf, or fern. At some stage in this degree the plant photosynthesizes and replenishes its nutrient reserves within the crown for the subsequent years harvest. In less warm districts, the asparagus fern will 'die off 'at some stage in iciness in similar style to a deciduous tree. The fern is then mulched into the soil or removed in coaching for harvest of the brand new asparagus spears in spring. The fern is not removed until completely 'dead' so that the asparagus root or crown is replenished as nutrients are reabsorbed from the stem and foliage. This step is followed by a shallow cultivation previous to spear emergence to incorporate the trash, smooth up weeds and provide a clear floor for harvesting.

Cultivar Selection

Potential growers will need to source an asparagus variety that has marketplace acceptability, is suitable in your nearby climate and, ideally, and if feasible, is raised organically (that is a demand of the national trendy for natural and bio dynamic produce). A few sorts will be greater suited to processing; others to the clean market. Will you grow green or white asparagus? Many european sorts are incorrect for warmer inland and northern districts because the spears tend to lose satisfactory because the leaf bracts open in advance within the hotter weather. Studies has shown that sure californian varieties are satisfactory acceptable below those conditions. Select varieties which have been bred for disease resistance, specially *fusarium sp.* and *phytopthera sp.* Resistance. Pre-planting and establishment status quo is the vital growth level for asparagus. Seedling transplants are not unusual, even though those are fairly sluggish to set up and full production can not be anticipated for at least 3 years. Crowns are more dependable and may yield in advance depending on crown length. The young crowns are commonly planted in a shallow trench that is progressively crammed in as the crown matures. This leaves the asparagus growing on a low mound (the mound is higher for white asparagus).

Crop Nutrients and Soil Control

Asparagus grows in a wide variety of soils however grows excellent in a light well-drained soil with a excessive nutrient content material. Deep alluvial soils offer gold standard situations for increase. However, those soils may be liable to erosion, low in natural matter and may be leached of important nutrients. The soil have to be analysed for its nutrient fame well before planting to determine pre-planting nutrient necessities and a fertility software must be applied as a minimum 2 years earlier than planting. If a green manure crop is to be grown as part of the fertility program it'll require incorporation nicely before planting in order that organic residues are well decomposed. Compost is an notable soil amendment and must be taken into consideration an critical addition preplanting, with annual programs for the hooked up asparagus stand. An incorporated green manure crop can supply from eighty to one hundred fifty kg real n in line with yr, whilst composted cow manure (analysed at 2 per cent n on a non-dry weight basis), implemented at 6 tonne/hectare/annum, should supply one hundred twenty kg actual n. Because of variability among batches of compost every ought to be separately analysed for nutrient content material. If externally sourced, you ought to also check for heavy metals. Phosphorus should be implemented as reactive phosphate rock (rpr) a season prior to planting as well as under the flowers at planting and in each year of production. Rpr ought to be checked for cadmium. Stages need to no longer exceed 20 ppm – mg/kg in fertilizers or manures. Potassium can be provided organically through composts and seaweed extracts. Commercial natural fertilizers that may supply most vitamins are available, however these ought to be carefully. Evaluated for value and effectiveness. A few natural certification organizations have accredited quite a number business natural fertilizers. Reliance on those, as opposed to correct soil control practices (inexperienced manuring and use of compost) is discouraged.

Weed Management

Weed manage is important in the course of asparagus status quo and is first-class started properly earlier than planting. This includes selecting an area with few weeds and the use of cultivation and sowing activities to encourage preplanting germination of weed seeds. Inexperienced manure and cereal crops planted previous to asparagus will help to out-compete weeds in addition to including to soil natural remember and fertility tiers. Difficult to control weeds may also want to be removed by means of hand. Weed management following planting can be tough because the younger asparagus plant lacks vigour and the competitiveness of a mature stand. Mulch can be implemented around seedlings.

Intercropping

An asparagus crop will now not attain complete productiveness for at least three years, so returns might be tremendously low at some stage in the status quo years. The put off in monetary go back can be in part offset by using intercropping among the asparagus rows with other greens or herbs. Planting an intercrop also facilitates reduce weed problems in the inter-row region. Any intercrop requires precise nutrition and water management to lessen opposition with the asparagus, and care is needed to ensure the crop does no longer host pests or illnesses of asparagus. Intercropping of asparagus with a low-developing leguminous cover crop may be powerful in lowering inter-row weeds, providing vitamins to the asparagus and enhancing soil shape. Studies has shown that there may be a competitive effect from cowl crops during the status quo yr and the first few years following transplanting the asparagus. Offering sufficient nutrients and water to the quilt crop may additionally reduce this effect. Two capability cowl cropping structures may be adopted. The primary kind involves planting a low growing perennial cover crop, including clover, this is mulched or slashed regularly for max gain. A side-throw mulcher, slasher, or instead a forage harvester might be used, with the cuttings directed across the asparagus flowers. This affords a nutrients and mulch benefit to the asparagus plant life. If the use of this system, enough distance need to be retained among the asparagus rows to allow for equipment operations which includes slashing of the asparagus fern in iciness, cultivation of weeds and harvest arrangements. Cultivation over the crop row need to be shallow or prevented altogether to avoid destructive the crowns. Soil might also need to be hilled up over rows to growth intensity. An opportunity cover cropping system entails sowing an annual cowl crop following elimination of the fern in iciness and incorporating it as a green manure or slashing/mulching it previous to spear emergence within the spring.

Harvesting

The harvest season for asparagus extends from august in hotter districts via to December in cooler districts. Harvesting has been prolonged the use of crop manipulation techniques including 'mom fern tradition'. In this machine, one spear is authorized to proceed to fern, whilst different spears inside the crown remain harvested. The 'mom fern' gives a few supplementary vitamins, barely prolonging the harvest. This method of extending the harvest, however, may also have an effect on crown durability and, therefore, the full efficient capacity of the asparagus stand. Harvesting is finished in the course of the early morning when it's mild cool.

Disease Management

1. Purple Spot (*Stemphylium vesicarium/Pleospora herbarum*)

Symptoms

Sunken, purple, oval-shaped lesions that develop on asparagus spears. Epidemics may affect 60-90 per cent of the spears. Tan to brown lesions on the fern, including the needle like leaves (cladophylls). May expand, coalesce and cause defoliation.

Management

☆ Scout the fields for symptoms of the disease.

- ☆ Use the TOM-CAST disease forecaster to predict when to apply fungicides. Refer to MSU Extension bulletin E312 for fungicides recommended for purple spot.
- ☆ Controlling purple spot enhances fern vigor and may aid in managing soilborne pathogens.
- ☆ Burn crop debris in late fall or winter – not feasible in large acreage due to human and environmental safety concerns.
- ☆ Till to bury crop debris – leads to wind erosion, reduced spear quality and marketability, and damage to crowns and roots.

2. Rust (*Puccinia asparagi*)

Symptoms

- ☆ Foliar disease.
- ☆ Four different lesions caused by four spore types.
  1. **Spring:** Oval, light-green lesions form.
  2. **Early summer:** These lesions become sunken and turn orange.
  3. **Mid-late summer:** New lesions on the foliage are brick-red colored and appear "blistered" above the plant surface.
  4. **Early fall:** Lesions on the foliage become black.
- ☆ Severe infections can defoliate plants.

Management

- ☆ Scout for aeciospores on young asparagus or early volunteer plants in early summer.
- ☆ Plant crop rows with ample spacing and in the direction of the prevailing winds to increase air movement and minimize periods of prolonged foliar wetness.
- ☆ Plant asparagus cultivars that are moderately disease resistant or tolerant (slow rusting).
- ☆ Eliminate asparagus seedlings and volunteers. Destroy abandoned fields.
- ☆ Controlling rust enhances fern vigor and may aid in managing soilborne pathogens.
- ☆ Apply fungicides. Refer to E312 bulletin for fungicides recommended for rust.
- ☆ Burn crop debris in late fall or winter – not feasible in large acreage due to human and environmental safety concerns.
- ☆ Till to bury crop debris – leads to wind erosion, reduced spear quality and marketability, and damage to crowns and roots.

3. Fusarium Crown, Root and Lower Stem Rot (*Fusarium oxysporum* f. sp. *asparagi Fusarium proliferatum*)

Symptoms

- ✰ Infection more likely when plants are stressed by drought.
- ✰ Russet-colored lesions on roots, lower stems and/or crowns.
- ✰ Yellowing of fern of infected plants.
- ✰ Damping-off of seedlings in crown nurseries.
- ✰ Crown death in nursery and commercial fields.
- ✰ Can shorten the lifespan of production fields by 50 per cent despite good cultural practices. One of the causes of declining fields.

Management

- ✰ Irrigate fields during times of drought.
- ✰ Fumigate fields the fall before planting. Refer to MSU Extension bulletin E312 for fumigants recommended for Fusarium.
- ✰ Avoid rotations to corn and planting on previous asparagus fields for a minimum of four years.
- ✰ Avoid tillage that may damage crown and roots.
- ✰ Select vigorous cultivars, avoid overharvesting and control pests (weeds, diseases, insects).

4. Phytophthora Crown, Root and Spear Rot (*Phytophthora asparagi*)

Symptoms

- ✰ Infection more likely when soils are wet.
- ✰ Spear rot begins as soft, water soaked lesions and/or shriveling occurring slightly above or below the soil line. Continued growth of infected spears result in shepherd's crook of spears and fern.
- ✰ Infected crowns and roots show water-soaked lesions and/or shriveling, but the tissue remains firm at the lesion site.
- ✰ Yellowing of fern of infected plants.
- ✰ Crown death in nursery and commercial fields.
- ✰ Can shorten the lifespan of production fields by 50 per cent despite good cultural practices. One of the causes of declining fields.

Management

- ✰ Apply fungicides. Refer to MSU Extension Bulletin E312 for fungicides recommended for Phytophthora.
- ✰ Avoid planting into low-lying or poorly drained fields.

Insect Pest Management

1. Asparagus Miner (*Ophiomyia simplex* Loew)

Family–*Agromyzidae*

Symptoms:

- ☆ Small black fly less than 0.2 inches long.
- ☆ Characteristic mining damage along stem base: discolored reddish brown tunnel in a serpentine pattern.
- ☆ Pupae are brownish red, about 0.2 inches long.
- ☆ Maggots are white, eel-like (early instars) or plump (later stages).

Management

- ☆ Use degree-day model on MSU's Enviro-weather website to properly time foliar insecticide sprays for peak adult abundance in the field.
- ☆ Remove field debris over the winter to prevent buildup of pest, and destroy debris or transport it far from asparagus production fields.
- ☆ Remove volunteer asparagus plants around fields.

3. Common Asparagus Beetle (*Crioceris asparagi*, L.)

Family–*Chrysomelidae*

Symptoms

- ☆ Adults are a 0.25 inches long with straight antennae.
- ☆ Adults have few black spots on a red thorax (middle segment), the abdomen (main body) has a patchy creamy yellow coloration interspersed with black and a maroon border around the rim.
- ☆ Larvae are muck-grayish to off-white with small, black heads and small, stout legs. Eggs are small, oval and grayish black: stuck to fern forming straight rows in groups of three or more and pointing away from the plant.

Management

- ☆ Use scouting for presence of adults and larvae in the afternoon since this is the time of greatest activity. Sampling at other times of day will result in an underestimation of the actual abundance of the beetle.
- ☆ During the harvest season, apply insecticides when 1-2 per cent of the spears have eggs or damage. During the fern stage, apply sprays when 50-75 per cent of the plants are infested with larvae or when 5-10 per cent of the plants are infested with adults.

- ☆ The thresholds for fields with repeated outbreaks of asparagus beetles in previous years may be lower since asparagus beetle damage can be cumulative.

3. Spotted Asparagus Beetle (*Criocerisduo decimpunctata*, L.)

Family–*Chrysomelidae*

Symptoms

- ☆ Adults are bright orange with six small black spots located on each wing cover.
- ☆ Adults are slightly larger than common asparagus beetles.
- ☆ Larvae are grayish white.

Management

- ☆ Remove volunteer asparagus plants.
- ☆ Spotted asparagus beetles are not damaging as a larva, and are much less of a problem as an adult than common asparagus beetles.

4. Asparagus Aphid (*Brachycorynella asparagi*)

Family–*Aphididae*

Symptoms

- ☆ Adults and nymphs are bluish gray, milky opaque with a powdery texture. They are some of the smallest aphids at about 0.04-0.0625 inches, lacking visible cornicles, or paired tubes extending from the rear part of the abdomen on the sides.
- ☆ Both winged and wingless forms may be present on the plants.

Management

- ☆ Remove volunteer asparagus plants.
- ☆ Monitor fields one to two times per week after harvest by beating the plants over a white surface, then inspect for aphids. Do this on multiple plants and locations across the field.
- ☆ An insecticide application may be warranted when 5 per cent of ferns show asparagus aphid damage.

5. Japanese Beetle (*Popillia japonica*) Newman

Family–*Scarabaeidae*

Symptoms

- ☆ Adults are usually 1/3 of an inch long with metallic, emerald green head and thorax and bronze wing coloration.
- ☆ Grubs are creamy-white, C-shaped and consume roots of grasses.

Management

- ✰ Pheromone traps are commercially available and help in detecting the beginning of flight.
- ✰ Scout by scanning the top of the asparagus fern to check for aggregations of beetles.
- ✰ Target adults with a registered insecticide when populations build up.

CONCLUSION

Concluded that the organic production of asparagus is most fruitful for growers. Growers gets more return from the crops. Using less input and get high output. Organically produced asparagus is beneficial for the human health and preventing the soil fertility and ecosystem also.

REFERENCES

Asparagus Breeding Program (Rutgers) https: //breeding.rutgers.edu/ asparagus/

Commercial Asparagus Production, HO-66 (University of Kentucky, 2008) http: //www.ca.uky.edu/agc/pubs/ho/ho66/ho66.pdf.Kentucky Department of Agriculture Division of Value-added Plant Production: Organic Program (KDA) https: //www.kyagr.com/marketing/organicmarketing. html. Sample Asparagus Production Budget for Kentucky (University of Kentucky, 2016) https: //www.uky.edu/ccd/sites/www.uky.edu.ccd/ files/asparagus_budget_2016.pdf.

Environmental Science and Technology online. 11 January 2006. Organic vegetables not pesticide free. Science News.

National Organic Program (NOP) http: //www.ams.usda.gov/nop

Organic Weed Management (eOrganic, 2013) https: //eorganic.org/node/2551

Organic Asparagus Production (NCAT-ATTRA, 2001) https: //attra.ncat.org/ attra-pub/summaries/summary.php?pub=377

Organic Weed Control Toolbox (eOrganic, 2019) https: //eorganic.org/ node/2782

Resource Guide to Organic and Sustainable Vegetable Production (NCAT-ATTRA, 2012) https: //attra.ncat.org/attra-pub/summaries/summary. php?pub=19.

Vegetable Production Guide for Commercial Growers, ID-36; includes Organic Manures and Fertilizers: Appendix C (University of Kentucky) http: // www2.ca.uky.edu/agcomm/pubs/id/id36/id36. pdf

Chapter 15

Organic Mushroom Production

Ajendra Kumar[1]*, D. Ram*[2] *and Ranjeet Kumar*[1]

[1]*Research Scholar,* [2]*Professor,*
Department of Plant Pathology, Acharya Narendra Deva University of Agriculture and Technology Kumarganj, Ayodhya–224229, Uttar Pradesh

ABSTRACT

Mushrooms are the edible fungus presently gaining the popularity in the Indian Kitchen due to its high nutritional value in economical rates. The mushroom industry in India is growing at steady rate because of less production cost and higher returns. Mushroom cultivation acquire very less area as compared to other field crops Mushroom plays an important role in the ecology of the forest, they decompose dead plants, including dead trees, branches, leaves, fruits, seeds, and animal excreta on the ground. Inorganic soil, 90 per cent of dead plants are made up of wood with cellulose and lignin, which makes them decay very slowly. Mushrooms produce an enzyme to decompose these substances more rapidly and create nutrients for other plants and microorganisms, thus completing the natural forest growth cycle. In recent days people of the cities and villages also are much of awared about the consequences of excess use of chemicals in the cultivation therefore they are moving towards the organically produced agriculture food material. Organic Mushroom is the best option for the better health and nutrition. Presently there are three kind of mushroom generally consumed by us namely white or button mushroom, oyster mushroom and paddy straw mushrooms.

Keywords: *Mushroom, Button mushroom, Oyster, Paddy straw, Growth cycle.*

INTRODUCTION

Mushroom production is an indoor activity under a controlled environment and without the use of soil as growing media. It is grown on compost, raw biomass, or wood. Mushroom is a fruiting body of micro-organisms called fungi. They form a mycelium in their stage of growth, which generates spore for dispersal in the gills: Generally lack chlorophyll, hence they don't synthesize their food through photosynthesis.

In Asia continents the environment is much suitable to grow different types of mushroom which are not only grown in low cost invested but also giving good returns, positively contributing to the farmer's economy and ultimately to the economy of our country. Before starting mushroom cultivation for commercial purposes, it's important to understand the different varieties of mushrooms, their attributes, qualities, environmental needs, and characteristics of growth. Different mushrooms have different market demands and profit margins, as well as growing challenges. It's also important to be aware of potential problems, such as pests, disease, and inclement weather.

Nutritional Composition of Mushroom

It is known as the 'meat' of the vegetable world. Since they were discovered, slowly and gradually, mushrooms are now used extensively in cooking in many cuisines, notably Chinese, Korean, European, and Japanese. With a wide variety of mushrooms, the possibilities and recipes are endless and with the different varieties available in the market, Mushroom owes their popularity not to just their delectable taste but their composition and nutritional profiles. While all edible mushrooms provide you with essential nutrients, below mentioned are some mushrooms that are nutrient dense and packed with antioxidants that may help you live a healthy life. Let us get to know the mushrooms and their benefits.

Mushrooms contain protein, vitamins, minerals, and antioxidants. These can have various health benefits. For example, antioxidants are chemicals that help the body eliminate free radicals. Free radicals are toxic by-products of metabolism and other bodily processes. They can accumulate in the body, and if too many collect, oxidative stress can result. This can harm the body's cells and may lead to various health conditions. Among the antioxidant agents in mushrooms are: Selenium, vitamin C, choline.

Gruen and Wong (1982) indicated that edible mushrooms were highly nutritional and compared favorably with meat, egg and milk food sources. Crisan and Sands (1978) observed that mushrooms in general contain 90 per cent water, 10 per cent dry matter with the protein content varying between 27 and 48 per cent and carbohydrates are less than 60 per cent and lipids are between 2 to 8 per cent. Orgundana and Fagade (1981) indicated that an average mushroom is about 16.5 per cent dry matter out of which 7.4 per cent is crude fiber, 14.6 per cent is crude protein and 4.48 per cent is fat and oil.

Table 15.1: Nutient Potential of Mushrooms

Nutrient	*Amount of Nutrient (in one cup of mushrooms)*	*Recommended Daily Intake*
Energy (calories)	21.1	1,600–3,200
Protein (g)	3.0	46–56
Carbohydrate (g)	3.1, including 1.9 g of sugar	130
Calcium (mg)	2.9	1,000–1,300
Iron (mg)	0.5	8–18
Magnesium (mg)	8.6	310–420
Phosphorus (mg)	82.6	700–1,250
Potassium (mg)	305	4,700
Sodium (mg)	4.8	2,300
Zinc (mg)	0.5	8–11
Copper (mcg)	305	890–900
Selenium (mcg)	8.9	55
Vitamin C (mg)	2.0	65–90
Vitamin D (mg)	0.2	15
Folate (mcg DFE)	16.3	400
Choline (mg)	16.6	400–550
Niacin (mg)	3.5	14–16

Organic Mushroom Cultivation Procedure

Mushroom cultivation in India is gaining popularity very rapidly. In current times it is considered as most profitable business in least investment and cultivation area. Because of its high nutritional values it gain attraction to health conscious person. In recent days people of the cities and villages also are much of aware about the consequences of excess use of chemicals in the cultivation therefore they are moving towards the organically produced agriculture food material. Organic Mushroom is the best option for the better health and nutrition. Presently there are three kind of mushroom generally consumed by us namely white or button mushroom, oyster mushroom and paddy straw mushrooms. In these three White or Button mushroom are the most popular and largely cultivated in India because of less expenditure. For the growing of button mushroom an organic compost is prepared. For effective mushroom production each grower must product economically and efficiently as possible as the highest quality of mushrooms. This can be achieved by using best strains which must be high yielding, visually attractive and having desirable flavour and resistance to adverse climate and pest and disease. The Brown variety is found superior on the white as it can tolerate the adverse condition better than white mushroom.

Organic Compost

In mushroom cultivation the most is the most requisite thing, the basic component of organic compost are horse dung, straw (wheat), gypsum and poultry manure. The horse dung should not be mixed with that of other animals and should be freshly collected then mixed with finely chopped straws of wheat. The formed mixture has to even spread in the composting yard followed by sprinkling the mixture with water so as to wet the straws. The mixture is then heaped and kept like that for fermentation. Once you smell ammonia from the heap, that means the compost has opened. The heap has to be turned in every 3 days and the same has to be sprinkled with water. On the 3rd or 4th time of turning, around 25 kilos of gypsum has to be added for each ton of compost.

Spawn

It is the propagating material used for planting beds by the cultivatiors. The spawn quality directly resonates with the quality of mushroom to be produced. Another basic ingredient that matters in cultivation is the compost that is used for growth of the mushrooms. Few other factors that matters the most include containment temperature, humidity (ranging between 85 to 95 per cent), ventilation and sterilization (by pasteurization at 37 °C).

First step begins with a compost yard in an open space. Concrete yards have to be prepared on raised platforms. Raised platforms won't let water accumulate into the heaps of compost. One more thing to consider is to cover the compost in the open so as to avoid rain spillage over the same.

Tray Filling of Compost

The compost becomes dark brown with no smell or trace of ammonia whenever it is ready for spawning. Also the pH levels become neutral and the water content should be moderate before you put the compost in the trays. If the compost is dry, the same has to be sprayed with a bit of water to make it moist. If the water content is more, then the excess water has to be evaporated. The tray has to be at least 15 cm deep and the size has to be 100 x 50 x 15 cm. Softwood should be used for the trays having pegs in 4 corners so that they can be stacked on top of one another leaving sufficient space between two trays for the various operations. The trays are filled with lightly compressed and levelled compost.

Buying Organic Spawn

Spawn are the propagating material as a fungal tissue for mushroom. Plug spawn is applied for long cultivation. The seeds are the mixture of compost. The utensils are to be washed with formalin solution along with your hands before handling the seedlings. The add seed in the compost @ 0.5 to 0.75 per cent.

The composting process shall ensure that the substrate has reached a temperature of at least 65 °C for about 6-7 days prior to use. All composts and growing media used (from the commencement of the composting process) shall be audited and verified for compliance with this Standard by the accredited certification body.

Casing Soil

The pH levels for the casing soil has to maintained within 7.5 - 8.5 and sanitized before application soil loaded on the cemented ground and treated with 4 per cent formalin; then covered with polythene and make air tight so that the vapours removed from the formalin completely kill the pathogen as well as their spore and soil sanitized. Logs, sawdust or other wood based material when used as substrate shall come from wood, trees or logs that have not been treated with prohibited substances.

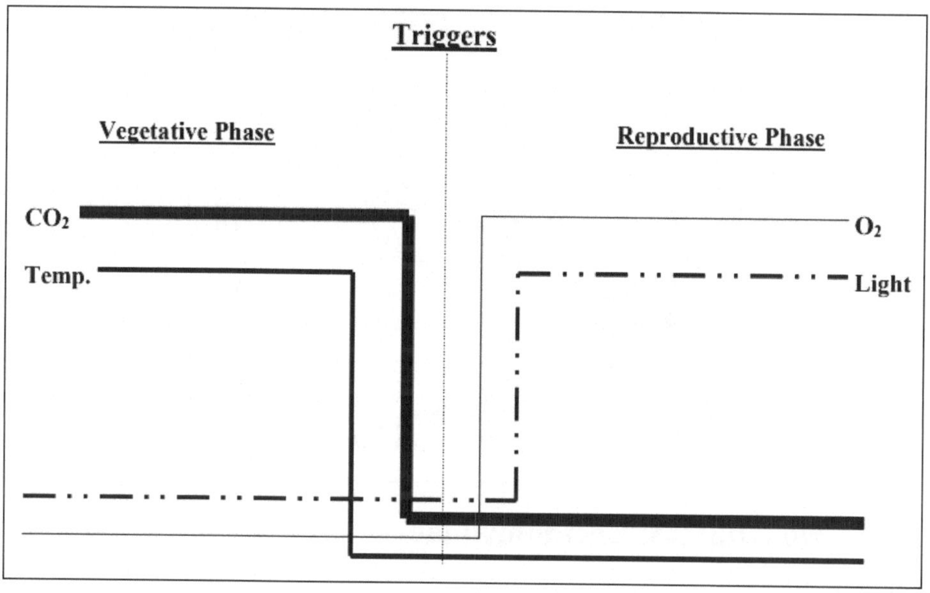

A 3-4 cm thick soil is layer on compost once the surface is covered with white mycelium of this fungus. There should be proper ventilation and spraying of water to maintain the humidity in growing chamber has to be maintained.

Pin Development

Once mycelium starts growing onto the surface of soil cooling down is done. After 5-6 days of cooling down due to colder air and lower CO2 the fluffy mycelium starts to contract resulting in the formation of a pin or pinhead. This period is called pin formation.

Cultivation of mushroom requires cold temperatures and best grown in autumn season. The temperature has to be gradually reduced from 23 °C till 17 °C over a period of 4 days. Reductions in the temperature once the mycelium grow completely. The lowering of temperature triggers the mycelium into sprouting the mushroom. Little buds tend to form which later develop into mushrooms. These little buds called pins and air temperature and humidity plays a major part in its growth.

Pest and Disease Control in Organic Mushroom Cultivation

Major pests include Cecid fly, Phorid fly, Sciarid fly, Nematodes. Proper sanitation should be followed in mushroom farming through pasteurization of the soil medium before a fresh crop is grown. Pasteurization kills the insects, nematodes, pest fungi and any other pests that might be living on the inside the compost.

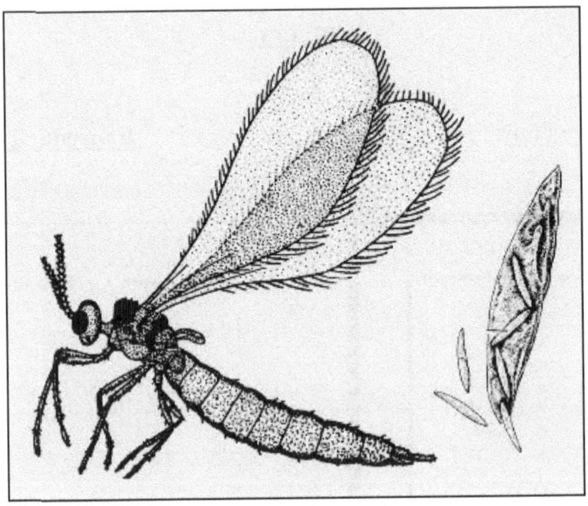

Cecid Fly (Diptera: Cecidomyiidae)

Identification

Cecid larvae are legless maggots bluntly pointed at both the ends uually white and orange in colour spread through out the mushroom. The adults of cecid flies are rarely seen because under most conditions, larvae become 'mother larvae' giving birth directly to 10-30 daughter larvae. This species usually does not become a pupa and subsequent adult that must mate before laying eggs. (Source- American mushroom institute).

Reproduction is accomplished without mating and gives rise to daughter larvae directly (paedogenesis). When conditions are optimal, this method of reproduction can result in very rapid multiplication of this pest, leading to very large number of larvae, tens of thousands per square metre.

(*Source*: Bug guide.Net)

Springtail (Colembolla Entomobryidae)

Identification

Springtails are tiny, 1 mm long, silver coloured, stout, having long tails. They are so named because they have forked structure or furculum with which they jump. The furculum arises on the ventral side of the fourth abdominal segment. During rest period, the furculum is folded forward under the abdomen.

The adults are of ground colour with light violet band along sides of the body. Intensely dark and rounded scales are present all over the body. The mouth parts are elongate and stylet like and are concealed within the head. Antennae are short and with a few segments.

Both nymphs and adults feed on mycelium and destroy planted spawn. They also attack stalks and caps. They are found in straw, compost and casing layer. Oyster mushroom is most preferred.

Control

1. Follow proper sanitation and pasteurization.
2. Crop should be raised above the ground level.
3. Never use spent compost.
4. Spray malathion 50EC at 0.05 per cent on the infested place.
5. Use diazinon 20EC @ 15 ml in 100 kg wheat straw compost.

Nitidulid Beetle (Coleoptera: Nitidulidae)

Identification

This is a serious pest of oyster mushroom. Female lays eggs singly between the gills. Grubs are the damaging stage which feed between gills and then on stapes of mushroom, making they hollow.

Control

1. Over matured mushroom should not be left unharvested.
2. Apply bleaching powder to repel these beetles.

Mushroom Nematodes, *Ditylenchus* spp. (Tylenchida: Anguinidae) and *Aphelenchoides* spp. (Aphelenehida: Aphelenchoididae)

Nematode infestation in mushroom is unique in a way that the mushroom nematodes have not only adapted themselves fully with ecological requirements of the crop and multiply very fast to occur in high number in mushroom beds, but these are only examples of ectoparasitic forms, capable of inflicting 100 per cent crop losses.

These are small microscopic thread like organisms which can be found in almost any type of environment, all types of soils and areas, where organic matter is found and are capable of rapid multiplication. The myceliophagous nematodes are the ones which suck the mycelial sap and leave it devitalized.

Control

1. Maintain strict hygienic conditions in the growing rooms by providing doors fitted with fly proof wire mesh and complete sanitation around the mushroom farm.
2. The peat heat temperature during pasteurization must be maintained at 58-59 °C for four hours.
3. The casing soil should be sterilized properly either with steam heat (65 °C) or with formation.
4. Wooden trays should be sterilized properly either with steam, formalin or dipped in hot water (80 °C) for 2-3 minutes.
5. Growing of nematode-resistant strains in rotation with the crop of common white button mushroom helps to reduce the inoculum level of the nematodes in the mushroom house.
6. The nematode trapping fungi like *Arthrohotrys* spp. can be used as biological control agents for controlling mushroom nematodes. The fungus traps the nearby nematodes and kills them.

Harvesting of Organic Mushroom

Harvest the mushroom by twisting them carefully at the soil level or by using sharp knife. Pulling can damage other surrounding mushrooms. The beds will continue to produce for upto about 6 months. The temperature is kept steady at 18 °C. Mushroom grow best at this temperature. They will grow 3 cm in a week, which is usual size of harvesting.

Mushroom can be either harvested by hand or the machine but hand picking of mushroom ensure better quality because mushroom removed from the beds undamaged. 9 days after 1st flush and 2nd flush will be harvested. The 2nd flush often consist of larger, but fewer mushroom than the first flush. After the 2nd flush of mushroom has been picked, the cell need to be cleaned. First a fall the cell is pasturized with steam to kill any remaining fungus to ensure that there is no transfer from cycle to cycle. During steam cleaning the temperature in the cells reaches 70° C for 8 hours. After the steam pasturizatrion the compost is removed from the from the beds and then it is thoroughly cleaned one more time and then it is ready to be filled again.

Registration and Management of Organic Mushroom Farming Site

The farmer has to submit an organic management plan to ACB (Accredited Certification Body) in order to register his or her farm. In return the ACB will verify the organic management plan and inspect the farm before giving the certification. The plan has to be updated yearly.

Any sort of non-organic farming has to remain separated from the organic farming site. Be it farming facility wise or any other machinery, tools, containers *etc.* Even the compost used in the process has to be organic in nature.

Organic Mushroom Production

REFERENCES

Chang, S. T. and P. G. Miles (1989). Edible Mushrooms and Their Cultivation. CRC Press, Inc., Boca Raton, FL. 345 P.

Hayes, W. A. and S. T. Chang (1978). The Biology and Cultivation of Edible Mushroom. Academic Press. New York, NY

Jayachandran, M., Xiao, J. and Xu, B. (2017). A critical review on health promoting benefits of edible mushrooms through gut microbiota. In *International Journal of Molecular Sciences* (Vol. 18, Issue 9). https: //doi. org/10.3390/ijms18091934

Krieger, L.C. (2010). The Mushroom Handbook. Sufi Press. 578 P. ISBN 486-21861-9

Oei, P. (1991). Manual on Mushroom Cultivation: Techniques, Species and Opportunities for Commercial Application in Developing Countries. TOOL Foundation, Amsterdam, The Netherlands. 249 P.

Rambelli, A. and U. G. Menini (1983). Manual on Mushroom Cultivation. FAO, Plant Production and Protection paper, No 43. UN. Rome, Italy.

Swayne, J.B. (1950). Handbook of Mushroom Culture. Kennett Square, PA 1934

Zadrazil, F. (1974). The Ecology and Industrial Production of *Pleurotus ostreatus, Pleurotus florida, Pleurotus cornucopiae,* and *Pleurotus eryngii*. The Mushroom Research Institute, Japan.

Chapter 16

Preparation Methods of Composting and Vermicomposting in Organic Farming

Shweta Chaturvedi* and D. Ram

Department of Fruit Science, Acharya Narendra Deva University of Agriculture and Technology, Kumarganj, Ayodhya, Uttar Pradesh
**e-mail:shwetachaturvedi2622@gmail.com*

ABSTRACT

Under the influence of chemical intensive systems, horticulture crops have seen significant increases in production, productivity, and area expansion during the last two decades. Unbalanced use of chemical fertilizers, particularly nitrogenous fertilizers, has had negative environmental consequences in some areas, contaminating soil and ground water resources. Soil quality particularly that of organic matter and micronutrient deficiencies, is deteriorating, posing a threat to food security by threatening the sustainability and quality of produce. Furthermore, indiscriminate use of agro-inputs, particularly chemical fertilizers and pesticides, has resulted in the development of pesticide resistance in pests, as well as the irreversible destruction of beneficial species such as honeybees, pollinators, parasitoids, and predators, as well as harmful pesticide residues in the end product, which has a negative impact on productivity and food safety. On-farm production of high-quality organic inputs should be prioritized for cost-effective and long-term production. Only a few organic farming approaches use locally accessible materials to produce all inputs, such as compost and vermicompost.

Keywords: *Organic inputs, Compost, Vermicompost.*

INTRODUCTION

Following the green revolution, agrochemicals were first used in commercial horticultural crop farming. In many locations, the indiscriminate use of agricultural chemicals has transformed the entire production environment. As a result, the number of pests has increased dramatically, and there is now a widespread perception that using agricultural chemicals is not the best solution for long-term horticulture crop production. A growing awareness of environmental conservation as well as the health risks posed by agrochemicals has resulted in a significant shift in consumer preference toward food quality, particularly in industrialized countries. The presence of organic matter in the soil is critical for preserving soil fertility and minimizing nutrient losses. Compost is an organic fertilizer that enriches the soil with organic materials and nutrients. A chemical fertilizer may be necessary to quickly supply a crop with the nutrients it requires. Chemical fertilizers, in contrast to organic fertilizers, aid plants right away; organic manures must first be broken down into nutrients (by soil organisms) before they can be used by plants. Chemical fertilizers, on the other hand, are depleted by the end of the season, whereas organic matter continues to improve soil fertility and structure as well as water storage capacity. Furthermore, the presence of organic matter guarantees that the crop makes better use of the chemical fertilizer. Plant nutrients are retained by organic matter, which prevents fertilizer from being washed away. Applying chemical fertilizer to a soil that is deficient in organic matter is a waste of money if it is not done in conjunction with steps to enhance the amount of organic matter in the soil.

Preparation of Compost

Man can control and speed up the natural decomposition process in the soil. Organic waste is collected and placed in a heap if possible. Because the heap is almost entirely made up of organic materials, the decomposition process is more intensive and the conditions are more favourable. The final output is decomposed organic matter containing humus and nutrients. This is termed as compost. Compost is an organic fertilizer that can be incorporated into the soil.

A good composting process goes through three stages in order to be effective:

- ☆ A Phase of Heating (Fermentation)
- ☆ A Cool Down Phase
- ☆ A Maturation Phase

Heating Phase

The compost heap begins to heat up significantly during the initial stage of composting. The breaking down of the organic matter's complex and tough fibrous structure causes this reaction, which is known as fermentation. The fermentation (decomposition) process is most active near the heap's centre. A lot of elements are necessary to get the fermentation process started rapidly and successfully. First and foremost, the compost heap should be made up of various organic components. Second, the appropriate bacteria must be present. Finally, it is critical that there is enough oxygen and water. Heat is generated quickly if these three requirements are met. Fermentation causes microorganisms to proliferate and alter rapidly, speeding up the heating process. This initiates a self-accelerated process. After 4-5 days, the fermentation stage begins and can last up to 1-2 weeks. In the compost heap, maximum fermentation occurs at a temperature of 60-70 degrees Celsius. The essential bacteria may perish if the temperature is too high, and decomposition will come to a halt.

Cooling Down Phase

The fermentation phase gradually gives way to the chilling phase. Decomposition takes place without much heat being generated, and the temperature gradually declines. New species of microorganisms turn the organic components into humus at this time. The temperature reduces from 50 degrees centigrade to 30 degrees centigrade inside the heap, which stays clammy and steamy. The process can be quickened or slowed down by adjusting the temperature, air, and water supply. The length of time it takes for the heap to cool down is determined by the type of material used, the amount of care paid to it, the climate, and other factors. The cooling down period normally lasts a few months, but it might take up to a year in extreme cases.

Maturation Phase

The temperature drops to 15-25 degrees Celsius in this final stage of decomposition, depending on the region. Apart from the aforementioned microorganisms, the extensive soil fauna is also active at this period. Earthworms, in particular, feed on strongly degraded organic material in temperate zones, contributing to decomposition. Termites, in particular, perform a crucial role in tropical and semi-arid climates, yet they can also be highly bothersome. This phase never truly ends; the decomposition process might continue at a sluggish rate indefinitely. If the compost is crumbly and looks like nice brown/black organic soil, it's ready to utilize.

Methods of Composting

There are many ways of making compost.

Indore Method

The Indore Method is much used for composting in layers.

Construction of Heap

Twigs and cane shoots should form the foundation of the heap. On top of that, the following layers are stacked:

- ☆ A layer of stiff organic material that takes a long time to degrade;
- ☆ A 2 cm layer of animal manure, compost, or biogas tank sludge
- ☆ A thin layer of earth; the soil should be taken from the top (moist) soil layer (top 10 cm) (*e.g.* from under trees). This ensures that the appropriate microorganisms are added to the heap. This layering process is repeated until the heap reaches 1.5 to 2 metres in height. As a result, the heap is made up of several levels. The heap should be constructed quickly, ideally within a week.

Turning Over

The heap must be turned over on a frequent basis during decomposition in order to keep it well aerated and ensure that all of the material is converted into compost. After 2 to 3 weeks, the heap should be turned over for the first time. The heap is dismantled and rebuilt next to the original heap. The layers are jumbled together, and the heap is turned inside out and upside down. A base of coarse plant material is laid down once more. The drier and less decomposed outside portion of the old heap is then placed in the centre of the new heap. Before the heap can be built up any more, the drier material will need to be irrigated. The rest of the material is used to cover the core. After 3 weeks, the heap is turned over a second time, and it may be necessary to flip it over a third time. After each turning over operation, repeat the moisture and temperature tests a few days later.

Time for Decomposition

Decomposition is complete if the plant material has changed into an unrecognizable crumbly, dark mass. Twigs and thick stems do not decompose completely and can still be seen. The breakdown process in the Indore Method takes 3 months under ideal conditions, but it might take up to 6 months under poor conditions.

Banglore Method

Another prominent composting process is the Bangalore Method. The heap is built in the same way as the Indore Method. In less than a week, a multi-layered compost heap is created.

The following are the differences between it and the Indore Method:

The heap is completely covered with mud or grass sods a few days after it are finished, effectively sealing it off from the outside air. Organic material continues to decompose, but now other types of microorganisms are keeping the process going. The substance is decomposed much more slowly by these microorganisms. As a result, it takes longer for compost to form than using the Indore Method, even though the compost quality is comparable.

Heating Process Method

The Bangalore Method is very similar to this method. The procedure, on the other hand, is unique and may be used to compost huge amounts of organic material.

A Continuous Heap System

A continuous heap system is used in the Heating Process Method. That is, new heaps of organic materials are constantly being created, heaped up, and treated. On the first day, a heap of all available materials is created. This heap has a minimum ground surface area of 1 x 1 metre and a maximum ground surface area of 3 x 3 metre, with a height of roughly 1 metre. Allow two days for the heap to settle.

Decomposition begins on its own within the heap. The air is driven out of the stack by trampling over it after these two days. The heap has been deprived of so much oxygen that it resembles the Bangalore Method's covered heap in appearance. On the fourth day, a new heap is constructed on top of the previous one. The first heap is completely sealed off from the outside air by this new heap. Every day, a fresh heap is created using the Heating Process Method. In fact, a new heap is created adjacent to the first one on the second and third days. A heap is constructed on top of the first one on the fourth day. As a result, on day 5, a fresh heap is added to the second one.

Trench Composting

Trench composting is similar to pit composting, except that instead of pulling the compost out of the pit and spreading it on land, plants are grown directly onto the trench. A trench should be dug initially. The size is determined by the amount of material available and the number of plants to be planted in the trench. It can have a width of 50cm to several metres, a depth of lm or less, and a length of any length. After that, fill it out as follows:

1. 10cm of difficult-to-decompose material (stalks or agricultural leftovers)
2. 10cm of easily decomposable material (fruit and vegetable leftovers)
3. Add 2cm of animal dung to the mix (if available)
4. To collect the microorganisms required for the composting process, a thin layer of soil from the surface of arable land is removed.
5. Repeat these layers until the pile is about 50cm above the ground
6. Continue layering until the pile is about 50cm above ground level.
7. Cover with soil, grass, or leaves (such as banana leaves) to minimise water and nutrient loss, and set aside for one month before planting.

Basket Composting

If you don't have enough composting materials, you can still make good use of them by composting in a basket. It's extremely useful in home gardens for growing food. It should be done in the following ways:

1. Dig a 60-cm-diameter, 60-cm-deep circular hole.
2. Line the bottom with hard-to-decompose material (twigs, stalks).
3. Add 8 cm of animal manure
4. Place 15 cm of green vegetation (new leaves with high water content) on top of the soil.
5. Pour in 0.5 cm of ash
6. Fill the hole with grass or leaves to avoid water and nutrient loss.
7. Cover with grass or leaves to prevent water and nutrient loss.
8. Make a circular outline of the hole using a spherical basket, 10cm in height, by weaving thin sticks together.

Production Technology of Vermicompost

Selection of Suitable Earthworm

Vermicomposting is a method of composting that involves the employment of earthworms, which eat digested biomass and excreta. Vermicompost is the common name for this type of compost. Earthworms that can be found locally are also utilised for vermicomposting, however their feeding mode is very slow, and the earthworm that lives below the soil is not suited for vermicomposting. Red worms (*Eisenia foetida*) and African earthworms (*Eudrillu sengenae*) are two promising vermicomposting worms. For vermicompost production, the two worms can be combined together.

Selection of Site

Vermicompost can be made everywhere there is shade, high humidity, and a cool temperature. Unused buildings, such as an abandoned cow or poultry shed, can be repurposed. If it is to be made outdoors, a shaded location is chosen. To protect the procedure from direct sunlight and rain, a thatched roof could be installed. Waste piled up for vermicomposting.

Method for Vermicomposting

1. Vermicomposting of wastes in field pit
2. Vermicomposting of wastes on ground heap method
3. Vermicomposting of wastes on improved method

Procedure of Vermicomposting

1. Compost can be made in a concrete tank (the size of which is determined by the availability of raw materials).
2. Collect and breakdown the weed biomass in the sun for 7-10 days or until it is completely decomposed. Chop the necessary hard materials.
3. Pour slurry of cow dung on the mound to help it disintegrate quickly.
4. Cover the bottom of the tank with a thin layer of surface soil/sand (1-2 inch).
5. Cover the soil or sand layer with fine bedding material such as partially decomposed cow dung/dried leaves, *etc.*
6. In the tank, layer the chopped bio-waste and partially decomposed cow dung up to a depth of 0.5-1.0 ft.
7. Cover the mixture with 1000-2000 worms/m ^{2} of any of the above earthworm species.
8. Use dry straw, thatch, or a gunny bag to cover the compost mixture.
9. Add water as needed to keep the moisture content between 70 and 80 percent.
10. Provide shade to shield the compost mixture from rain and intense sunlight.
11. When 80-98 percent of bio waste has decomposed, stop sprinkling water. The creation of granular structure in the compost at the tank's surface could be used to determine maturity.
12. Scrape the vermicompost layer by layer off the top of the tank and store it in the shade.

CONCLUSION

Compost and vermicompost if properly made can enhance the fertility and productivity of the soil. It will also maintain the soil health and will readily convert unavailable form of nutrients into available form. Hence, it should be practised by the farmers for long term benefits.

REFERENCES

Brinton, FW. (2000). Compost quality standards and guidelines, Woods End Research Laboratory, Inc. pp. 1-42.

Bansal, S. and Kapoor, KK. (2000). Vermicomposting of crop residues and cattle dung with *Eisenia foetida*. Bioresource Technology, 73(2), pp. 95-98.

Bitton, G. (2003). Wastewater microbiology. 3th Edition, John Wiley and Sons, Inc., pp. 247.

Cook, SMF. and Linden, DR. (1996). Effect of food type and placement on earthworm (*Aporrectodeatu berculata*) burrowing and soil turnover, Biology and fertility of soil, 21(3), pp. 201-206.

Dickerson, G. W. (2001). Vermicomposting, Extension Horticulture Specialist, Guide H-164, Cooperative Extension Service College of Agriculture and Home Economics, pp. 1-4.

Chapter 17

Zero Budget Natural Farming for Future

B.K. Prasad[1], Rajendra Kumar[2], G. Singh[3], R.K. Singh[1] and Om Prakash[4]

[1]Department of Genetics and Plant Breeding,
[4]Department of Animal Husbandry and Dairying,
Amar Singh College, Lakhaoti, Bulandshahar–203407, Uttar Pradesh
[2]Amar Singh College, Lakhaoti, Bulandshahar–203407, Uttar Pradesh
[3]Department of Genetics and Plant Breeding, P.G. College, Ghazipur, Uttar Pradesh

ABSTRACT

In the future, agricultural production is most likely to increase from providing a better environment through input management, and from crop varieties of fully exploiting the prevailing environment. However, the above achievement was driven by costly inputs, such as chemical fertilizers, seeds and pesticides and extensive irrigation. The rising cost of these external inputs will be a leading cause of indebtedness and suicide among farmers, while the impact of chemicals on the environment and on long term fertility will also be devastating. An alternative to the above farming system, zero budget natural farming is a type of farming that promotes low cost inputs such as the use of cow dung, aged cow urine, jaggery, pulse flour and other plant-based extract. In this farming system no external fertilizers and pesticides are used, hence they reduce the cost of cultivation by reducing dependency on external inputs like seeds, fertilizer, pesticides, etc., enhance soil fertility, enhance crop yields, reduce risks of insect pests and diseases, and protect from uncertainties of climate change by promoting the adoption of an agro ecology framework.

Keywords: *Zero budget natural farming, Organic farming, Jeevamrutha, Beejamrutha, Mulching.*

INTRODUCTION

The agricultural food production in India had increased from merely 50.82 million tonnes in 1950-51 to 230.67 million tonnes in 2007-08. This represents an increase of over 350 per cent during a period of 57 years. During this period, there was an increase of 27.87 per cent in the cultivated area, while yield per hectare increased by 255 per cent. As a result, the nation became almost self sufficient in food grains. But the population in India is growing at an alarming rate of around 2.5 per cent per year. This makes it necessary that the food grains production should also increase at least at the same rate or even at a faster rate to improve the nutritional status of masses. The above increase in yield resulted from the quantum and better management of inputs, such as fertilizers, irrigation water, plant protection and cultural practices, and from improved crop varieties (Singh, 2012). It is doubtful that the net cropped area can be increased indefinitely, but there is still some scope through double and multiple cropping. Improved input management practices are yet to be fully exploited, and vast tracts of cultivated lands are very poorly managed. In the future, agricultural production is most likely to increase from providing a better environment through input management, and from crop varieties of fully exploiting the prevailing environment. However, the above achievement was driven by costly inputs, such as chemical fertilizers, seeds and pesticides and extensive irrigation. The rising cost of these external inputs will be a leading cause of indebtedness and suicide among farmers, while the impact of chemicals on the environment and on long-term fertility will also be devastating. In this connection, an alternative to the above farming system, natural farming need not to spend money on inputs, or take loans to buy them (Sarma, 2016). Natural farming is a system where the laws of nature are applied to agricultural practices. This method works along with the natural biodiversity of each farmed area, encouraging the complexity of living organisms, both plants, and animals that shape each particular ecosystem to thrive along with food plants. Natural farming is an ecological farming approach established by Masanobu Fukuoka (1913–2008), a Japanese farmer and philosopher, introduced in his 1975 book The One-Straw Revolution. There are many working models of natural farming all over the world, the zero budget natural farming (ZBNF) is the most popular model in India. This comprehensive, natural, and spiritual farming system was developed by an Indian agriculturist and Padma Shri recipient, Subash Palekar. The term zero budget means the zero cost of production of all crops. ZBNF guides the farmers in practicing sustainable farming that helps in retaining soil fertility to ensure chemical-free agriculture and ensure low cost of production (zero-cost). In the other words, ZBNF refers to the process of raising crops without using chemical fertilizers and pesticides or any other external materials. Instead, farmers use low-cost locally-sourced natural concoctions, inoculums and decoctions based on cow dung, cow urine, jaggery, lilac, green chillies and many other such natural ingredients (https//www.bussiness-standard.com).

This chapter is aimed to provide up to date information about various aspects of ZBNF.

Objectives of ZBNF

This climate-resilient agricultural method, which is different from organic farming, aims at promoting agroecology and adopting low-cost agriculture practice wherein all critical inputs like seeds, fertilizer, pesticides, *etc.*, are gathered from the field and nothing is introduced from outside. Under ZBNF, neither fertilizer nor pesticide is used and only 10 per cent of water is to be utilized for irrigation as compared to traditional farming techniques.

The basic concept of ZBNF is to ensure that profit from supplementary crops is enough to cover the cost of production of the main crop and it has the potential to enhance the income or profit margins for farmers who are often under distress owing to debt.

Principles of ZBNF

This farming system is based on following principles:

1. Plants get 98 per cent of their supply of nutrients from the air, water, and sunlight and the remaining 2 per cent can be fulfilled by good quality soil with plenty of friendly microorganisms (just like in forests and natural systems).
2. In this farming system, soil is always supposed to be covered with organic mulch, which creates humus and encourages the growth of friendly microorganisms.
3. This farming system requires cow dung and cow urine (gomutra) of Indian breed cow only. Desi cow is apparently the purest as far as the microbial content of cow dung and urine.
4. A farm made bio-culture named 'Jeevamrutha' is added to the soil instead of any fertilizers to improve microflora of soil. Jeevamrutha is derived from very little cow dung and cow urine of desi cow breed.
5. Natural, farm-made pesticides like dashparni ark and neem astra are used to control pests and diseases.
6. Weeds are considered essential and used as a living or dead mulch layer.

Components of ZBNF

On the basis of principles of ZBNF, components of ZBNF can be classified into following four groups:

1. Jivamrita/jeevamrutha
2. Bijamrita/beejamrutha

3. Acchadana (mulching) and
4. Whapasa (moisture).

Jivamrita/Jeevamrutha

It is prepared by mixing fresh desi cow dung and aged desi cow urine, jaggery, pulse flour, water and soil in a plastic drum and kept for fermentation. After a few days, the above prepared microbial culture can be sprayed in the field. This microbial culture adds nutrients to the soil, and acts as a catalytic agent to promote the activity of microorganisms and earthworms in the soil. About 200 litres of jeevamrutha should be sprayed twice a month per acre of land; after three years, the system is supposed to become self-sustaining. Only one cow is needed for 30 acres of land, according to Mr. Palekar, with the caveat that it must be a local Indian breed not an imported Jersey or Holstein.

Bijamrita/Beejamrutha

It is a mixture of neem leaves and pulp, tobacco and green chillies prepared for insect and pest management, that can be used to treat seeds.

Acchadana (Mulching)

It protects topsoil during cultivation and does not destroy it by tilling.

Whapasa (Moisture)

It is the condition where there are both air molecules and water molecules present in the soil. Thereby, providing water to maintain the required moisture-air balance.

Advantages of ZBNF

The ZBNF method also promotes soil aeration, minimal watering, intercropping, bunds and topsoil mulching and discourages intensive irrigation and deep ploughing. Palekar introduces the most common composting worm, the European red wiggler (Eisenia fetida) to Indian soils. He claims these worms absorb toxic metals and poison groundwater and soil.

Which are the States with Big Plans?

According to the Economic Survey, more than 1.6 lakh farmers are practicing the ZBNF in almost 1,000 villages using some form of state support, although the method's advocates claim more than 30 lakh practitioners overall. The original pioneer was Karnataka, where the ZBNF was adopted as a movement by a State farmers' association, the Karnataka Rajya Raitha Sangha. Large-scale training camps were organized to educate farmers in the method. According to a survey carried out in those early years, ZBNF farmers all owned small plots of land, had some access to irrigation and owned at least one cow of their own (Jebaraj, 2019, 2021).

In June 2018, Andhra Pradesh rolled out an ambitious plan to become India's first State to practise 100 per cent natural farming by 2024. It aims to phase out chemical farming over 80 lakh hectares of land, converting the State's 60 lakh farmers to ZBNF methods (Jebaraj, 2021).

Is the Budgetary Support Enough?

Despite the ZBNF buzz caused by the Budget speech, the Finance Minister did not actually announce any new funding to promote it. Last year, the Centre revised the norms for the Rashtriya Krishi Vikas Yojana- Remunerative Approaches for Agriculture and Allied sector Rejuvenation (RKVY-RAFTAAR), a flagship Green Revolution scheme with an allocation of 3,745 crore this year, and the Paramparagat Krishi Vikas Yojana, which has an allocation of 325 crore and is meant to promote organic farming (Annadurai and Palaniappa, 1997 Baphna, 1997, Dixit, 1997, Narayan, 2005, Prakash, 1997) and soil health. Under the revised guidelines, both Centrally-sponsored schemes now allow States to use their funds to promote the ZBNF, vedic farming (http://www.yogickheti.org/), natural farming, cow farming and a host of other traditional methods (Jebaraj, 2021).

Andhra Pradesh says it has utilized 249 crore from these schemes to promote the ZBNF over a two and a half year period. The State estimates it will need 17,000 crore to convert all of its 60 lakh farmers to the ZBNF over the next 10 years (Jebaraj, 2021). However, this is only a fraction of the spending on Central government subsidies for fertilizers, pesticides and mass irrigation that has driven the Green Revolution model.

What Lies Ahead ?

NITI Aayog has been among the foremost promoters of Mr. Palekar and the ZBNF method. However, its experts have also warned that multi-location studies are needed to scientifically validate the long-term impact and viability of the model before it can be scaled up and promoted country wide.

The Indian Council of Agricultural Research is studying the ZBNF methods practised by basmati and wheat farmers in Modipuram (Uttar Pradesh), Ludhiana (Punjab), Pantnagar (Uttarakhand) and Kurukshetra (Haryana), evaluating the impact on productivity, economics and soil health including soil organic carbon and soil fertility.

If found to be successful, an enabling institutional mechanism could be set up to promote the technology, NITI Aayog vice-chairman Rajiv Kumar has said. The Andhra Pradesh experience is also being monitored closely to judge the need for further public funding support.

Similarities between Natural and Organic Farming

1. Natural and organic both are chemical free and more or less poison free farming methods.

2. Both systems discourage farmers from using any chemical fertilizers, pesticides on plants and in all agricultural practices.
3. Both farming methods encourage farmers to use local breeds of seeds, and native varieties of vegetables, grains, pulses and other crops.
4. Organic and natural farming methods promote non chemical and homemade pest control methods.

Differences between Natural and Organic Farming

Sl.No.	*Organic Farming*	*Natural Farming*
1	In organic farming, organic fertilizers and manures like compost, vermicompost, cow dung manure, *etc.* are used and added to farmlands from external sources.	In natural farming, neither chemical nor organic fertilizers are added to the soil. In fact, no external fertilizers are added to soil or given to plants whatsoever.
2	Organic farming still requires basic agro practices like ploughing, tilting, mixing of manures, weeding, *etc.* to be performed.	In natural farming, decomposition of organic matter by microbes and earthworms is encouraged right on the soil surface itself, which gradually adds nutrition in the soil, over the period.
3	Organic farming is still expensive due to the requirement of bulk manures, and it has an ecological impact on surrounding environments; whereas, natural agriculture is an extremely low-cost farming method, completely molding with local biodiversity.	In natural farming, there is no ploghing, no tilting of soil and no fertilizers, and no wedding is done just the way it would be in natural ecosystems.

REFERENCES

Annadurai, K. and Palaniappan, S. P. (2018). Organic Farming: Theory and Practice. Scientific Publishers, India.

Baphna, P. D. (1997). My successful experiments in agri-horticulture. Proceedings of National Seminar on the Indian Model for Rural Development in the 21st Century. Organized by Mahamana Malaviya Mission, B. H. U. Unit, Varanasi, 51- 53.

Dixit, P. K. (1997). Some aspects of sustainability of agriculture and of organic farming. Proceedings of National Seminar on the Indian Model for Rural

Development in the 21st Century. Organized by Mahamana Malaviya Mission, B. H. U. Unit, Varanasi, 59-62.

Elayaraja, M. and Vijai, C. (2020). Organic farming in India: Benefits and Challenges. European Journal of Molecular and Clinical Medicine 7 (11) 3021-3029.

http: //www.yogickheti.org/

Jebaraj, P. (2019). What is zero budget natural farming ?. The Hindu. July 28, 2019.

Jebaraj, P. (2021). The push for Zero Budget Natural Farming. The Hindu. Dec. 16, 2021

Narayan, S. (2005).Organic Farming In India: Relevance, Problems and Constraints. Department of Economic Analysis and Research,

National Bonk for Agriculture and Rural Development, Mumbai, pp. 93.

Prakash, Bhartendu (1997). Vichar karen karbnic-krishi ke labhkari pahluo par. *Proceedings of National Seminar on the Indian Model for Rural Development in the 21st Century*. Organized by Mahamana Malaviya Mission, B. H. U. Unit, Varanasi, 21-22.

Sharma, Prasada (2016). Campaign to reduce use of chemical fertilizers, pesticides." The Hindu, May 28, 2016.

http: //bit.ly/1tpq0rT.

Singh, B. D. (2012). Plant Breeding Principles and Methods. Kalyani Publishers, New Delhi

Zero budget natural farming in India. 52 Profiles on Agroecology. https: //www.fao.org

What is zero-budget natural farming? https: //www.business-standard.com

Chapter 18

Organic Waste Management and Soil Health

A.P. Singh[1] *and R.P. Singh*[2]

[1]*Department of Agricultural Chemistry,* [2]*Department of Agronomy, Amar Singh College, Lakhaoti, Bulandshahar, Uttar Pradesh*

ABSTRACT

Agricultural waste means anything which is left over from the harvested crop after the removal of economic produce from crop plants. However, nothing is waste in agriculture; this depends on the situation and varies from person to person. Development of alternative fertilizer material like enrichment of organic manure through minerals and their effect on crop productivity and soil fertility build -up has potential which will give the benefit of applying organic as well as inorganic fertiliser increasing crop production and maintaining soil quality. The organic matter or manure like FYM, compost, vermicompost can improve food quality. Compost not only reduce the dependence on chemical fertilizers but also increase the soil organic matter, improve soil physical environment microbial biomass, soil microbial activity. Carbon, inactivate the soil born pathogen, thereby encourage the growth, crop quality as well as sustain higher productivity. In an agricultural field regularly using imbalanced fertilizer and lower no use of organic manure soil reduced potential productivity at a fixed level of output. It is a widespread problem and a major issue because of its adverse effect impact on agricultural productivity and sustainability. Unsustainable activities including inadequate soil conservation, minimum use of organic manure, imbalance fertilizer use and improper irrigation management have a devastating impact on land resources in arid and semi-arid regions. The soils of arid zone are very deficient in organic matter.

INTRODUCTION

Traditionally, before sowing of agricultural crop, at the time of preparation of the field the farmer treats the crop residue present in that field as complete waste and throws it out of the field or burns it in the field itself. Stubble burning also comes under this, in fact due to the practice of single farming system and indiscriminate use of chemical fertilizers and pesticides, farmers are often unaware of the important uses of left out parts of crops or residues like dry fodder of straw, paddy, sugarcane leaves, sorghum, maize *etc.*, fruit or vegetable peel, flowers, leaf, root and stalk *etc.* are known as crop residues, that is, leaving the main products of crop like cereals, fruits, vegetables, *etc.*, all other parts of crop are crop residues. These remnants of what we think of as garbage is used to increase the fertility of the land, as animal feed, as fuel *etc.* and in many other works. It is also used in industries.

About 600 million ton total crop residue is generated in our country every year and about one third of which is used in the field for fodder, fuel and other works. In some parts of Uttar Pradesh, Haryana, Punjab, Bihar *etc.*, the residues of paddy and sugarcane crop is burnt in the field itself, due to which the environment is polluted (soil or air pollution) and the soil organisms are also destroyed or reduced and become inactive.

Effect of Chemical Fertilizers

Inorganic fertilizers have a very important role in agricultural crop production, but due to their use, all the necessary nutrients for the soil are not fulfilled. Only main elements like N, P and K are supplied to the soil and the deficiency of other elements continues, in addition to this, due to their excessive use, the environment is also polluted.

Loss of soil fertility, pollution of soil water, reduction in water holding capacity of the soil, the formation of hard layer on the soil, continuous decrease in the number and activity of micro-organisms present in the soil due to the increase in soil temperature and all this led to a bad impact on physical and chemical structure of the soil also the excessive release of greenhouse gases like CO2, CH4 and N2O *etc.*, causes increase in soil, crop and environment pollution. This in turn affects the health of soil microorganisms as well as the health of human being. Due to the rise in pollution of the environment, more attention should be paid on alternative sources of these for the supply of nutrients also there is a need of research on organic waste management. Application of organic manure in soil through better nutrient recycling improves the physical properties of soil and maintains the crop system. Although, large quantities of manures are required to fulfil the nutrient need of the soil. Thus, keeping crop production, environment, health *etc.*, in mind a balanced mixture of organic and inorganic compost or fertilizer can be used.

Recycling of organic waste has gained significant importance in the recent past which is considered as the precious resource for soil health restoration besides its importance as a source of nutrient. About 500 million tonnes of Crop Residues are produced annually out of which 70 per cent is cereal residues (Biswas 2011) estimated 141 Mt of surplus (after satisfying competitive uses namely fodder, fuel, thatching, packaging *etc.*,) crop residues improve in soil organic carbon, moisture, earthworms and microbes, decreases evaporation losses, moderates soil temperature *etc.,*. Technologies and policy support for efficient crop residues management are the need of the day for sustainable crop production. With the emerging concern on large quantity of agricultural waste being produced the concept of waste management becomes one of the keys focuses of sustainable agricultural development. Thus, the appropriate management of agricultural waste assumes a great significance. One of the alternative options of utilization of these large quantities of nutrient rich agricultural waste is by converting them into value added products like compost and recycles them back to the field which have drawn the attention of scientists to reduce environmental pollution and increase efficiency of carbon and nutrient rich inputs for higher productivity.

Types of Agricultural Waste

The term "Agricultural waste" is used in a broad sense to include wastes from agriculture and allied sectors. A variety of agricultural waste which is available for recycling are:

1. Farm wastes: crop residue, tree waste and aquatic waste
2. Livestock wastes: cattle shed wastes, piggery and poultry wastes and
3. Agro industry wastes: press mud, coir pith from coconut pith from coconut industry, rice husk, paper and pulp wastes, oil cakes *etc.*

Crop Residues

Crop residues are the remnants of crop plants left after the harvest. Being easily accessible to the farmers for application on his land, crop residues have traditionally played an important role in maintaining soil productivity. The dominant uses of crop residues in our country are as fodder for cattle, fuel for cooking, thatch material for housing and as organic resource. On the basis of crop production levels, it is estimated that 10 major crops (rice, wheat, sorghum,

Table 18.1: Nutrient Content in Different Crop Residues

Crop Residue	*Nutrient Content (per cent)*		
	N	*P_2O_5*	*K_2O*
Wheat	0.47	0.17	1.16
Rice	0.60	0.19	0.138
Maize	0.51	0.17	1.34
Sugarcane	0.39	0.17	1.27
Pulses	1.61	0.50	1.74
Groundnut	1.61	0.22	1.36
Potato	0.51	0.20	1.01
Pearl Millet	0.44	0.15	1.15

pearl millet, barley, finger millet, sugarcane, potato tubers and pulses) of India generate about 31 2.5 Mt of crop residues that have nutrient potential of about 6.46 Mt of N, P, K.

Harmful Effects of Recklessly Burning Crop Residues

Despite enormous benefits of the crop residues, most of the crop residues are removed or burnt. The main reasons for burning crop residues in field include unavailability of labour, high cost in removing the residues and use of combines in rice wheat cropping system especially in the Indo Gangetic plains. The intensification of land use results in less time between crops for managing these residues, which can interfere with tillage and seeding operations for the next crop. The lack of alternative uses for crop residues and lack of appropriate mechanization to handle increasing quantities of residues have driven the farmers increasingly to burn crop residues as a method of disposal. About 93 Mt of crop residues are burnt on-farm in India. In Punjab, for instance, about 70 per cent of the total rice straw produced from combine harvest is burnt in open fields. Similarly, after combine-harvesting wheat, 70 per cent of the wheat straw is converted to *bhusa* for animal feed and the remaining is burnt in the field. Unlike wheat straw, rice straw is not used as a cattle-feed because of high silica and oxalate content in it. The crop residues could be incorporated directly into the soils or recycled back to agricultural field through composting, vermicomposting or enriched composting along with addition low-grade minerals like rock phosphate and waste mica to improve the quality of compost.

The burning of crop residues causes atmospheric pollution due to the emission of toxic gases like Methane, carbon dioxide or nitrous oxide that poses a threat to human and Ecosystem health. Due to crop residues burning, in addition to loss of entire amount of C, 80 per cent of N, 25 per cent of P, 50 per cent of S and 20 per cent of K present in straw is lost due to burning, it also pollutes the atmosphere. If the crop residues are incorporated or retained, the soil will be enriched, particularly with organic carbon and N. Heat from burning residues elevates soil temperature causing death of bacterial and fungal populations. However, the death is temporary as the microbes regenerate after few days. Repeated burning in the field, however, permanently diminishes the microbial population. Open-field burning of crop residuesis recognised as a major contributor to reduced air quality and human respiratory ailments, particularly in North-Western India, which represent major irrigated rice ecosystems in India.

Organic Matter

Organic matter is dead and decaying residues of plants and other living beings. They deposit on the surface of soil or mixed with by various processes. Several microorganisms work on this organic waste or residue and convert the organic to inorganic form. In this way the organic waste or residues of plants and other living beings (fresh or decomposed), cells of soil organisms and other synthesized various matter by them are included in organic matter of soil. Plants are the main source of organic matter in the soil. They mix organic waste by the help of their leaves, branches, stems and roots in a huge amount. Soil microorganisms and their residue are also a source of organic matter. But they are placed after plant residue. Microorganisms assimilate this organic waste and residue by processing it. And these microorganisms are also a source of organic matter in the soil. Organic matter improves the quality of soil physically, chemically and biologically. Organic matter is defined as a combination of three substances:

1. Live Biomass, including animals, microbes and plant roots.
2. Recognisable dead and decaying biological matter.
3. Humid substances, the heterogeneous polymers formed during decay and degradation of plant, animal and microbial biomass

Organic matter is the chief source of minerals returned to soil. It has also much water holding capacity. It can hold minerals as ions by absorption and binds clay particles into aggregates. It makes clay soil porous thus increasing aeration of soil and also causes easy percolation of water in soil. Soil organic matter (humus) represents the equilibrium between input, originating from primary photosynthetic production, and the derivative and resynthetic processes associated with soil organisms. The input is largely of carbohydrate, lignin compounds, fats and proteins with smaller quantities of free amino acids, alkanes, terpenoids, carotenoids, flavonoids, alkanoids, polyphenols, resins *etc.* Most of these as carbohydrate, cellulose *etc.* are easily decomposed and have short residence time in soil. Soluble sugars, lipids, proteins, amino acids, organic acids *etc.*, are minor constituents organic matter. However, the major constituents of humus are humic complexes which make the greatest proportion of it in most soils.

Organic matter to soil is contributed by surface living plants and animals (dead leaves, twigs, branches of plants and animal bodies) as well as by micro and macro-organisms living in the soil. In thick forests in each season there accumulates thousands of tons of litter on the soil surface. All dead, fresh organic matter fallen recently to the ground is called 'litter' and partially decomposed litter is called '**duff**'. Thus, initial input of organic matter is in the form of surface litter and underground dead root matter. But as a result of various soil processes, chiefly activities of microorganism in the soil, this organic matter (litter) is grossly modified, which is now known as 'humus or soil organic matter in current practice. Humus is dark, finely divided amorphous organic matter. The composition of the organic matter of soil determined by litter chemistry and also influenced by the nature of microbiological resynthesizes. The rate of humus formation and its accumulation are thus governed by nature of microorganism in decompositions, temperature, moisture, aeration and pH of soil.

Table 18.2: Characteristics of Bacteria, Actinomycetes and Fungi of Surface Soils

Characteristic	*Bacteria*	*Actinomycetes*	*Fungi*
Population	Most numerous	Intermediate	Least numerous
Biomass	1-2 tons/ha	0-2 tons/ha	Largest biomass (2-5 tons/ha)
Degree of branching	Slight	Filamentous, but some fragment to individual cells	Extensive filamentous forms
Cell wall	Murien, teichoic acid, and lipopolysaccharide	As in bacteria	Chitin or cellulase
Growth rate	Exponential	Cubic	Cubic
Growth in liquid culture	Turbid	Pellets	Pellets
Aerial mycelium	Absent	Present	Present
Fruiting bodies	Absent	Simple	Complex
N-Fixation	Present	Present	Absent
Oxygen relations	Aerobic, anaerobic	Mostly aerobic	Aerobic except yeast
Moisture stress	Least tolerant	Intermediate	Most tolerant
Optimum pH	6-8	6-8	6-8
Competitive pH	6-8	>8	>5
Competitiveness in soil	All soils	Dominate dry, high pH soils	Dominate low pH soils

Table 18.3: Examples of Important Heterotrophic Surface Soil Bacteria

Organism	*Characteristics*	*Function*
Streptomycetes	$Gram^+$, aerobic, filamentous	Produce geosmins, earthyodor and antibiotics
Bacillus	$Gram^+$, aerobic, spore former	Carbon cycling, production of insecticides and antibiotics
Clostridium	$Gram^+$, anaerobic, spore former	Carbon cycling, fermentation, toxin production
Alcaligeneseutrophus	$Gram^-$, aerobic	2,4-D degradation via plasmid PJP_4
Rhizobium	$Gram^-$, aerobic	Symbiotic, N-Fixation with legumes
Frankia	$Gram^+$, aerobic	Symbiotic N-Fixation with non- legumes
Agrobacterium	$Gram^+$, aerobic	Important plant pathogen, cause crown gall disease
Methylosinus	Aerobic	Methane oxidizers that can cometabolite trichloro ethene (TCE) using methane mono oxygenase

Table 18.4: Approximate Biomass of Major Component of the Biota in Typical Temperate Grassland Soil

Component of Soil Biota	*Biomass (tons/ha)*
Plantroot	*Up to 90 generally about 20*
Bacteria	*1-2*
Actinomycetes	*0-2*
Fungi	2-5
Protozoa	0-0.5
Nematodes	0-0.2
Earthworms	0-2.5
Other soil animals	0-0.5
Viruses	Negligible

Organic Matter of Content of the Soil

The importance of organic matter has been well recognised starting from the formation of soil (weathering of rocks) to building up of soil fertility for raising a healthy crop with higher productivity. Since organic matter is the residue of living organisms which took nutrients from some sources for their survival, therefore, it is obvious that organic substances would liberate plant nutrients in available form during their mineralization. Organic matter, after certain degree of decomposition, forms humus (a storehouse of various nutrients) which improves the soil structure, drainage, aeration, increases water holding, buffering capacity, influences the solubility of the minerals and supplies energy for the growth and development of microorganism. All these lead to building up of soil fertility. In general, the composition of plant tissue (mature and fully dried) ranges around the following values:

Carbohydrates

Sugars and starches	1-5 per cent
Hemi cellulose	10-28 per cent
Cellulose	20-50 per cent
Fats, waxes, tannins, *etc.*	1-8 per cent
Lignin	10-30 per cent
Protein (simple water soluble and crude protein)	1-15 per cent

Advantages of Enrichment of Compost

In general, the bulky organic manures like FYM and compost contains, on an average, 0.5 to 1.0 per cent N, 0.2 to 0.5 per cent P_2O_5 and 0.5 to 1.0 per cent K_2O. Thus, if 120 kg of N is to be supplied through these organic manures, one has to add 20-25 t of compost per ha, which is not feasible as limited quantity of organic matter are available in our country. Moreover, the costs of preparation, storage, transport and application of FYM or compost to soils are very high. The demerits of these bulky manures can be overcome to a certain extent through preparation of enriched compost by adding nitrogen, phosphorus and potassium either alone or in combination. Enrichment of organic manure through minerals has the following advantages:

- ✰ Offers a potential avenue for the efficient utilisation of low-grade materials such as rock phosphate and waste mica (a K-bearing mineral).
- ✰ Enrichment of organic manures offers a versatile scope of applications, have a well-balanced composition and offer a mild and long-lasting effect.
- ✰ For better results farmers should use both enrichment of organic manures and inorganic fertilizers to reduce the dependency on chemical fertilizers.
- ✰ It increases fertilizer nutrient use efficiency and maintains soil organic matter.
- ✰ Lesser problems in handling, storage and transportation.
- ✰ These are eco friendly and can be used safely.
- ✰ It is compatible with soil microbes, improves rhizosphere microflora and hence ensure fertility of the soil. Enrichment of organic manure improves the organic matter content in soil, helps in the improvement in soil texture, water holding capacity, soil aeration for better root development.
- ✰ These fertilizers build and maintain the soil structure and energize soil life.
- ✰ It prevents the nutrient losses due to immobilization by microorganisms during decomposition of organic residues and adsorption due to high cation exchange capacity of organic matter.
- ✰ These fertilizers reduce nutrient losses due to leaching.
- ✰ Another added advantage of enrichment of organic manure is their slow nutrient release and residual effect on subsequent crop yields. In their study the authors found that residual effect of poultry manure led to 33 per cent yield increase in Maize over that of NPK fertilizer. This aspect on residual effect of enrichment of organic manure is now being studied in a follow-up experiment.
- ✰ Enrichment of organic manures is more concentrated manure, thereby, it reduces the bulk to be handle per unit of material.

CONCLUSION

Converting agricultural waste into wealth is a viable option for enhancing farmers income. From above discussion it can be concluded that organic manure and fertilizers can be prepared using crop residues and these fertilizers thus prepared have higher N, P and K content than ordinary compost. Using innovative technologies as well as improved farm implements, the farm waste have broader applications in current scenario rather than the earlier times. Therefore, it is clearly visible that development of organic fertilizer could be an alternative and cost-effective technology to utilise agricultural waste and low-grade materials such as rock phosphate and waste mica in place of costly water-soluble P and K fertilizer for crop production, maintaining soil fertility and hence, enriching soil health.

REFERENCES

Biswas, D. R. 2018, Agricultural wastes and their composting for sustainable crop production and soil health, 2, 11-19.

Shiva Dhar, Palta, R.K. and Singh, J.P. 2018. Agricultural biomass recycling and bio composting technologies. 3: 20-28.

Biswas, D. R., Narayanasamy, G. Dutta, S. C., Singh, G., Begum, Mamta, Maiti, D., Mishra, A. and Basak, B. B. 2009, changes in nutrient status during preparation of enriched organo-mineral fertilizer using rice straw, low grade rock phosphate, waste mica and phosphate solubilizing microorganisms. Communications in soil science and Plant Analysis 40: 2285-2307.

NRAES 1992, on farm composting(Ed. Rynk, Robert) Natural Resource, Agriculture and Engineering Service, Cooperative Extension, Ithaca, NY.

Singh, P. Avdhesh, 2019, Organic matter decomposition cycles of matter in nature, 2019 3: 137-144

Chapter 19

Role of Organic Farming in Soil Inhabiting Microorganisms

Krishna Kumar[1]*, Popin Kumar[2], Ajay Kumar[3] and Abhimanyu[4]

[1]*Assistant Professor, School of Agriculture, Galgotias University, Greater Noida–201301, Uttar Pradesh*
[2]*Research Scholar, Department of Plant Pathology, Sardar Vallabh Bhai Patel University of Agriculture and Technology, Modipuram Meerut–250110, Uttar Pradesh*
[3]*Assistant Professor, Amar Singh College (CCS University Meerut), Lakhaoti, Bulandshahar–203407, Uttar Pradesh*
[4]*Krishi Vigyan Kendra, Farrukhabad (CSAUAT, Kanpur)–222 131, Uttar Pradesh*
**e-mail: kkplantpathology123@gmail.com*

ABSTRACT

Soil microorganisms classified as bacteria, actinomycetes, fungi, algae, protozoa and viruses. Each of these groups has different characteristics that define the organisms and different functions in the soil it lives in. Importantly, these organisms do not exist in isolation; they interact and these interactions influence soil fertility as much or more than the organism's individual activities. Fungi are unusual organisms, in that they are not plants or animals. The fungi help the plant by giving it needed nutrients and the fungi get carbohydrates from the plant, the same food that plants give to humans. Bacteria are anywhere from 100 million to one billion bacteria in just a teaspoon of moist, fertile soil. They are decomposers, eating dead plant material and organic waste. Microorganisms like both bacteria and fungi, and have characteristics linking them to both groups. They are often believed to be the missing evolutionary link between bacteria and fungi, but they have many more characteristics in common with bacteria than they do fungi.

Algae are present in most of the soils where moisture and sunlight are available. They are capable of photosynthesis, whereby they and obtain carbon dioxide from atmosphere and energy from sunlight and synthesise their own food.

Keywords: *Pathogens, Inhabiting, Soil, Microorganism, Organic.*

INTRODUCTION

India is an Agrarian country with around 60 per cent of its people directly or indirectly depends upon agriculture. In archaic time the practice of agriculture was considered to be a maximum service to the society and this practice was inter-twined in their convention and culture. Since organic farming techniques have the probable to improve soil fertility, soil structure and soil moisture retention capacity. Organic management provides solutions to the problems associated with degradation of drylands and desertification.

The increase in organic agriculture development and the need for further promoting it has been summed up very well by the 42ndReport of the Standing Committee on Agriculture of 14th Lok Sabha. "The Committee feels that R and D in Organic farming is an option that will make agriculture in India more profitable as organic farming is more environmentally sustainable. Hence, it has to be included as a priority area in agriculture. Implications of organic farming on national food security may also be analysed". The study revealed that organic farming, in spite of the reduction in crop productivity by 9.2 per cent, provided higher net profit to farmers by 22.0 per cent compared to conventional farming. However, there was an overall improvement in soil quality in terms of various parameters, *viz.* physical, chemical, biological properties, availability of macro and micro-nutrients, indicating an enhanced soil health

Organic farming is defined as a production system that avoids or largely eliminates the usage of synthetic compounded fertilizers, growth regulators, pesticides, and farm animal feed additives. To maximize feasible extent, the organic farming system depends on crop rotation, green manures, legumes, animal manures, crop residues, off-farming natural squanders, and aspect of biological pest control to preserve soil fertility and productivity to sustain plant and thereby curbing pests, diseases, and unwanted weeds. Organic farming methods are internationally regulated and legally executed by many countries, based in great part on the standards set by the International Federation of Organic Agriculture Movements (IFOAM), an international umbrella organization for organics established in 1972.

After the launch of green revolution in India, substantial growth in the output of food grains was achieved. This was achieved through the using of improved crop varieties and higher levels of inputs of plant foods and plant protection chemicals. Only there has been a rise in production, which was complete by the pecuniary value of soil health. The organic farming and ecological agriculture are one of the alternative agriculture systems to conquer the problems of soil degradation and declining soil fertility.

Mahatma Gandhi pioneered organic farming direct his constructional programmes in several locations in India. It was the aggressive promotion of the green revolution, which led to a reverse in the spread of the Gandhian movement's organic farming programme. There is evidence to suggest that small farmers, especially in the Third World, are more likely to use agricultural practices, such as crop rotations and mixed cropping, which are the essential building blocks of organic farming. The evidence further suggests that while conventional agriculture goes better with large holdings, organic farming functions better in small farms. Study by Gupta and Verma (1997), comparing grain production in organic conventional methods, observed that as farm size increases, the advantages of organic rotation become less visible. Further, the study reported that on a smaller scale, organic farming was more profitable.

Table 19.1: Rank (on basis of total area) under Organic Agriculture

Rank (Total area)	*Country*	*Area under Organic Agriculture (ha)*	*Percentage of Total Agriculture Land*	*Number of Organic Farms*
1	Australia	12,294,290	2.8	1550
2	China	2,300,000	0.4	1600
3	Argentina	2,220,489	1.7	1486
4	USA	1,620,351	0.5	8493
5	Italy	1,148,162	9.0	45115
6	Uruguay	930,965	6.1	630
7	Spain	926,390	3.7	17,214
8	Brazil	880,000	0.3	15,000
9	Germany	825,539	4.8	17,557
10	UK	604,571	3.8	4485
11	Canada	604,404	0.9	3571
13	India	528,171	0.3	44926
	World	30,418,261	0.65	718,744

Source: Ramesh *et al.*, 2010.

India ranks 10th in the world in terms of cultivated land under organic certifications. The certified area includes 15 per cent cultivable area with 0.72 million Hectare and rest 85 per cent (3.99 million Hectare) is forest and wild area for collection of minor forest produces. The total area under organic certification is 4.72 million Hectare. India produced around 1.24 million MT of certified organic products which includes all varieties of food products namely Sugarcane, Cotton, Basmati rice, Oil Seeds, Pulses, Fruits, Dry fruits, Spices, Tea, Vegetables, Coffee and their value added products. The production is not limited to the edible sector but also produces organic cotton fibre and functional food products *etc.* Among all the states, Madhya Pradesh has covered largest area under organic certification followed by Himachal Pradesh and Rajasthan.

Table 19.2: The Status of Organic Production in India (2013-2014)

Total area under certified organic cultivation	0.72 M ha
Forest and wild area for collection of minor forest produces	3.99 M ha
The total area under organic certification	4.7 M ha
Total production	1.24 million MT
Total quantity exported	177766 MT
Value of total export	1328.61 crores

Source: APEDA, 2014.

Role of Microorganisms in Organic Farming

Soil microorganisms can be classified as bacteria, actinomycetes, fungi, algae, protozoa and viruses. Each of these groups has different characteristics that define the organisms and different functions in the soil it lives in. Importantly, these organisms do not exist in isolation; they interact and these interactions influence soil fertility as much or more than the organism's individual activities.

Fungi

Fungi are unusual organisms, in that they are not plants or animals. They group themselves into fibrous strings called hyphae. The hyphae then form groups called mycelium which are less than 0.8mm wide but can get as long as several metres. They are helpful, but could also be harmful, to soil organisms. Fungi are helpful because they have the ability to break down nutrients those other organisms cannot. They then release them into the soil, and other organisms get to use them. Fungi can attach themselves to plant roots. Most plants grow much better when this happens. This is a beneficial relationship called mycorrhizal. The fungi help the plant by giving it needed nutrients and the fungi get carbohydrates from the plant, the same food that plants give to humans.

Bacteria

There are anywhere from 100 million to one billion bacteria in just a teaspoon of moist, fertile soil. They are decomposers, eating dead plant material and organic waste. By doing this, the bacteria release nutrients that other organisms could not access. The bacteria do this by changing the nutrients from inaccessible to usable forms. The process is essential in the nitrogen cycle.

Actinomycetes

Microorganisms like both bacteria and fungi,and have characteristics linking them to both groups. They are often believed to be the missing evolutionary link between bacteria and fungi, but they have many more characteristics in common with bacteria than they do fungi. Actinomycetes give soil its characteristic smell.

Algae

Algae are present in most of the soils where moisture and sunlight are available. Their number in the soil usually ranges from 100 to 10,000 per gram of soil. They are capable of photosynthesis, whereby they and obtain carbon dioxide from atmosphere and energy from sunlight and synthesise their own food.

Function of Algae in Soil

- ☆ Playing an important role in the maintenance of soil fertility, especially in tropical soils.
- ☆ Adding organic matter to soil when they die and thus increasing the amount of organic carbon in soil.
- ☆ Acting as a cementing agent by binding soil particles and thereby reducing and preventing soil erosion.
- ☆ Helping to increase the water retention capacity of soil for longer time periods.
- ☆ Liberating large quantities of oxygen in the soil environment through the process of photosynthesis and, thus, facilitating submerged aeration.
- ☆ Helping to check the loss of nitrates through leaching and drainage, especially in un-cropped soils.
- ☆ Helping in the weathering of rocks and the building up of soil structure.

Protozoa

They are colourless, single-celled animal-like organisms. They are larger than bacteria, varying from a few microns to a few millimetres. Their population in arable soil ranges from 10,000 to 100,000 per gram of soil and they are abundant in surface soil. They can withstand adverse soil conditions.

Function of Protozoa in Soil

- ☆ Protozoans derive their nutrition from feeding or ingesting soil bacteria and,thus, they play an important role in maintaining microbial/ bacterial equilibrium in the soil.
- ☆ Some protozoa have been recently used as biological control agents against organisms that cause harmful diseases in plants.

☆ Several soil protozoa cause diseases in human beings that are carried through water and other vectors.

Viruses

Soil viruses are of great importance, as they may influence the ecology of soil biological communities through both an ability to transfer genes from host to host and as a potential cause of microbial mortality. Consequently, viruses are major players in global cycles, influencing the turnover and concentration of nutrients and gases. To explore the role of the viruses in plant health and soil quality, studies are being conducted into virus diversity and abundance in different geographic areas (ecosystems). It has been found that viruses are highly abundant in all the areas studied so far, even in circumstances where bacterial populations differ significantly in the same environments. Soils probably harbour many novel viral species that, together, may represent a large reservoir of genetic diversity.

Nematodes

Not microorganisms (strictly speaking), nematode worms are typically 50 microns in diameter and one millimetre in length. Species responsible for plant diseases have received much attention, but far less is known about much of the nematode community, which play beneficial roles in soil. An incredible variety of nematodes have been found to function at several levels of the soil food web. Free-living nematodes can be divided into four broad groups based on their diet. Bacterial-feeders consume bacteria. Fungal-feeders feed by puncturing the cell walls of fungi and sucking out the internal contents.

Nematodes are important in mineralising, or releasing, nutrients in plant-available forms. When nematodes eat bacteria or fungi, ammonium is released because bacteria and fungi contain much more nitrogen than the nematodes require. Nematodes may also be useful indicators of soil quality because of their tremendous diversity and their participation in many functions at different levels of the soil food web.

Role and Functions of Microorganisms

☆ Soil microorganisms play an essential role in decomposing organic matter, cycling nutrients and fertilising the soil.

☆ Without the cycling of elements, the continuation of life on Earth would be impossible, since essential nutrients would rapidly be taken up by organisms and locked in a form that cannot be used by others.

☆ The reactions involved in elemental cycling are often chemical in nature, but biochemical reactions, those facilitated by organisms, also play an important part in the cycling of elements.

- Soil microbes are also important for the development of healthy soil structure. Soil microbes produce lots of gummy substances (polysaccharides and mucilage, for example) that help to cement soil aggregates.
- This cement makes aggregates less likely to crumble when exposed to water.
- Soil microorganisms are both components and producers of soil organic carbon, a substance that locks carbon into the soil for long periods.
- Abundant soil organic carbon improves soil fertility and water-retaining capacity.
- There is a growing body of research that supports the hypothesis that soil microorganisms, and fungi in particular, can be harnessed to draw carbon out of the atmosphere and sequester it in the soil.

CONCLUSION

It can be concluded that microorganisms like both bacteria and fungi, and have characteristics linking them to both groups. They are often believed to be the missing evolutionary link between bacteria and fungi, but they have many more characteristics in common with bacteria than they do fungi. Algae are present in most of the soils where moisture and sunlight are available. They are capable of photosynthesis, whereby they and obtain carbon dioxide from atmosphere and energy from sunlight and synthesise their own food.

REFERENCES

APEDA(2014). Agricultural and Processed food product Export Development Authority, Ministry of Commerce and Industries, GOI, www.apeda.gov.in.

Archna, K. (2013). Role of Indian Government in the Development of Organic Agriculture, *IOSR Journal of Agriculture and Veterinary Science* (IOSR-JAVS), e-ISSN: 2319-2380, p-ISSN: 2319-2372.2 (6): 32-39.

Balasubramanian, A., Siddaramappa, R. and Rangaswami, G. (1972). "Effect of organic manuring on the activities of the enzymes hydrolysing sucrose and urea and on soil aggregation, " Plant and Soil, 37(2): 319-328.

Bhattacharyva, P. and Chakraborty, G. (2005). Current status of organic farming in India and other Countries, *Indian Journal of Fertilisers*, 1(9): 111-123.

Brook, and Bhagat, G. (2004).Organic by default: The irony of organic farming in India, available at: http: //www.ecoworld.org/Home/Articles2.cfm?TID=349.

Chhonkar P.K. (2002). "Organic farming myth and reality, " in Proceedings of the FAI Seminar on Fertilizer and Agriculture Meeting the Challenges, New Delhi, India.

Clark, M.S., Horwath, W., Shennan, R.C. and Scow, K.M. (1998). "Changes in soil chemical properties resulting from organic and low-input farming practices, " *Agronomy Journal*, 90(5): 662-671.

Doran, J.W. and Zeiss, M. (2000). "Soil Health and Sustainability: Managing the Biotic Component of Soil Quality." *Applied Soil Ecology* 15 (1): 3-11.

Ettema, Christien, H. (1998). "Soil Nematode Diversity: Species Coexistence and Ecosystem Function." *Journal of Nematology*, 30(2): 159-69.

Gaur A.C. and Prasad, S.K. (1970). "Effect of organic matter and inorganic fertilizers on plant parasitic nematodes, " *Indian Journal of Entomology*, (32): 186-188.

Gaur, A.C., Sadasivam, K.V., Vimal, O.P. Mathur, R.S. and Kavimandan, S.K. (1973). "Studies on the humification of organic matter in a red Rakar soil" Zentralblatt fur Bakteriologie, 128(1): 149-161.

Ikerd J. (1993).Two related but distinctly different concepts: organic farming and sustainable agriculture. Small Farm Today, 10(1): 30-31.

Kulhade, A., Gupta, A.D., Mishra, A. and Singh, S.R.K. (2016). Role of organic farming in Indian agriculture. *Flora and fauna*, 22(1): 41-48.

Meena, R.K., Meena, R.S., Naik, B.S.S.S., Meena, B.L. and Meena, S.C. (2020). Organic farming-concept, principles, goals and as a sustainable agriculture: A review, *International Journal of Chemical Studies*, SP-8(4): 24-32.

Meena, R.K., Neupane, M.P. and Singh, S.P. (2015). Effect of phosphorus levels and bio organic sources on growth and yield of rice (*Oryza sativa* L.). *International journal of Agricultural Science*, 11(2): 26-289

Minhas, R.S. and Sood, A. (1994). "Effect of inorganic and organic on yield and nutrients uptake by three crops in rotation in aid alfisol, " *Journal of the Indian Society of Soil Science*, 42: 27-26.

Pandey, J. and Singh, A. (2012). Opportunities and constraints in organic farming. An Indian perspective.*Journal of Scientific Research*, 56: 47-72.

Ramesh, P., Panwar, N.R., Singh, A.B., Ramana, S., Yadav, S.K., Shirvastava, R. and Rao, A.S. (2010). Status of organic farming in India, *Current Science*, 98(6): 1190-1194.

Ranganathan, D.S. and Selvaseelan, D.A. (1997). "Mushroom spent rice straw compost and composted coir pith as organic manures for rice, " *Journal of the Indian Society of Soil Science*, 45(3): 510-514.

Rigby, D. and Aceres, D. (2001). Organic farming and the sustainability of agricultural systems. *Agricultural Systems*, 68: 21-40.

Sharma, A.R. and Mitra, B.N. (1990). "Complementary effect of organic material in rice-wheat crop sequence, " *The Indian Journal of Agricultural Sciences*, 60(3): 163-168.

Singh, G., Singh, O.P., Yadava, R.A. Singh, P.P. and Nayak, R. (1998). "Response of rice (*Oryza sativa*) varieties to nitrogen levels in flash flood conditions, " *Indian Journal of Agronomy*, (43)3: 506-510.

Singh, G.B. and Dwivedi, B.S. (1996). Integrated nutrient management for sustainability. *Indian farming*, 46(9): 9-15.

Yadav, S.K., Babu, S., Yadav, M.K., Singh, K. and Yadav, G.S. (2013). A Review of Organic Farming for Sustainable Agriculture in Northern India. *International Journal of Agronomy*, (718145): 8.

Chapter 20

Insect Pest Management in Organic Farming

***Rajanish Chandra Mishra*[1], *Anuj Kumar Gautam*[2] *and Rajkumari*[3]**

[1]Subject Matter Specialist (Plant Protection), [2]Subject Matter Specialist (Animal Science) and [3]Subject Matter Specialist (Home Science),
Krishi Vigyan Kendra, Jalaun, Banda University of Agriculture and Technology, Banda, Uttar Pradesh
e-mail: kaushal4980@gmail.com

ABSTRACT

Over-reliance on the use of synthetic pesticides in pest management programs causes pest resurgence, pest resistance to pesticides, lethal and sub-lethal effects on non-target organisms, including humans and also pollutes the environment. To combat these side effect organic farming is emerging as a viable option that avoids or largely excludes the use of synthetically produced agricultural inputs like fertilizers, pesticides,growth regulators, livestock feed additives etc. made even more complex when factoring in insect and disease pests. Since the use of synthetic pesticides is prohibited, the organic cropping system should be focused on the prevention of pest outbreaks rather than coping with them after they occur. No single method is likely to be adequate for all pests. Successful pest management in organic farming depends on the incorporation of a number of management strategies viz., cultural, mechanical, physical, biological and use of plant extracts etc.

INTRODUCTION

Although insect-pests, weeds and pathogens are thought to destroy 10-40 per cent of the world's gross agricultural production, pesticides are not the perfect option for controlling insect-pests and pathogens. Indiscriminate use of synthetic pesticides in crop protection programs has resulted in disturbances to the environment, pest resurgence, pest resistance to pesticides, pest out break and lethal and sub-lethal effects on non-target organisms, including humans. These side effects have raised public concern about the routine use and safety of pesticides. During the last two decades, there has been a significant sensitization of the global community towards environmental preservation and assuring of food quality. The promoters of organic farming consider that it can meet both these demands and become the mean for complete development of rural areas. After almost a century of development, organic agriculture is now being embraced by the mainstream and shows great promise commercially, socially and environmentally. According to the IFOAM,organic agriculture is guided by four principles: **Health** (soil, plant, animal, and human), **Ecology** (living ecological systems and cycles), Fairness (environment and life opportunities), and Care (protect the health and well-being of current and future generations as well as the environment).

Organic Pest Management offer approaches that reduce reliance on pesticides. Pest management in organic farming is achieved by using appropriate cropping techniques, biological control, and natural pesticides (mainly extracted from plant or animal origins). Insect- pests problem in organic farming is managed by using integration of cultural, mechanic, physical, and biological methods.

Pest Management Practices in Organic Farming

Pests in a crop do not automatically result in damage or yield loss. In some instances, low levels of pest feeding have been shown to increase crop yields. Once infestation levels reach a certain point, however, they can produce economic losses. Thresholds vary with the crop and the pest in question and must be closely monitored by the producer. Planning for effective insect and disease management must involve the entire farm operation and use all information available. Practices used in organic farming are based on the three management strategies, which include prevention, monitoring, and suppression. These practices will be intensively discussed in the following:

1. Monitoring and Identification of Insect-Pests and Diseases

Monitoring is the regular inspection or scouting of field crops for pests, including insects, pathogens, nematodes, and weeds, to determine their abundance and level of damage. It serves as an early warning system for the presence of pests and diseases providing information for decision-making regarding management action and evaluation of control methods. Insect pests can be monitored through visual observation, pheromone and light traps, sticky traps, water traps, yellow traps, sweep nets, beating trays, and pitfall traps. Scouting data are used to develop economic thresholds, a useful decision-making tool to start control action when a pest population reaches or exceeds the specified economic threshold.

Identification of insect pests and their natural enemies is an important step in any pest management program. Insect pests and natural enemies could be identified by using keys and field guides or otherwise consulting an official identification bodies. Unlike insect pests, plant pathogens including fungi, bacteria, virus, and nematodes are difficult to identify in the field and may need laboratory diagnosis. However, signs of insect damage and symptoms of plant diseases may be easily distinguished in the field. Weeds could be easily identified using key and field guides.

2. Tactics Used for Pest Prevention and Suppression in Organic Farming

A successful pest management program in organic farming incorporates a variety of pest management tactics such as cultural, mechanical/physical, biological, and biopesticidal (allowed for organic use) tactics individually or in combination. Each control tactic, discussed below, employs a different set of mechanisms for preventing and suppressing pest populations.

2.1 Cultural Pest Control

Slight modification in the cultural practices will have an impact on the whole ecosystem. These practices can enhance the agricultural biodiversity and thus have a greater role to play in the management of insect pests as well as the pathogens. These are following types:

A. *Sanitation and Crop Residue Destruction*

This method is used to reduce pest infestation through the removal of breeding and hibernating sites. It involves eradication of harmful weed hosts or alternate hosts, timely destruction of crop residues, cleaning of field borders of alternate hosts, and removal of scrub or shelter in which pests might hide, *e.g.*, in orchards, destruction of cull or dropped fruits and pupation sites for codling moth, apple maggot, and plum curculio may be achieved by means of livestock, and suction or cultivation equipment.

B. *Tillage*

Tillage destroys insects overwintering in the soil as eggs, pupae, or adults, and reduces pest problems. Organic producers usually rely on tillage to control weeds and to prepare the soil for planting. However, cutworm and slug problems can also increase where tillage is reduced.

C. *Use of Resistant Cultivars*

Plant breeders traditionally have placed more focus on creating pests-resistant varieties. However, insect resistant varieties are an effective defense. Even when insect resistant cultivars are not available, some varieties may be less attractive to pest species or tolerate more damage than others. For example, onion crop with wider leaf angle and round leaves support less number of thrips incidence (Patil *et al.*, 1988).Moreover, plants with glossy foliage are more resistant against onion thrips (Molenaar, 1984). Although, resistant varieties and natural controls generally works together to suppress insect-pests population.

D. Time of Sowing and Planting

This is used largely to avoid invasion by migrants, or the oviposition period of particular pests, and the introduction of disease in the crop by insect vectors to synchronize the pest attack with its natural enemies, with weather conditions that are adverse for the pest, or with the abundance of an alternative host; to make it possible to decoys the crop before the pest enters diapause. Timing sowing used to allow young plants to establish a tolerant stage before pest attack, to reduce the susceptible period of attack, to mature the crop before a pest becomes abundant, to allow it to compensate for damage and to fill gaps where plants have been damaged or killed, and to avoid the egg-laying period for a particular pest.

E. Plant Density

If increasing the population density of a crop increases beneficial insects, it can lead to a decrease in pest insects. In some crops, close row spacing proliferate control by beneficial insects.

F. Fertility Management

Plant nutrition can influence the feeding, longevity and fecundity of phytophagous pests; the common fertilizer elements (nitrogen, phosphorous and potassium) have direct and indirect effects on pest suppression. In general, nitrogen in high concentrations has the reputation of increasing pest incidence, particularly of sucking pests such as mites and aphids. On the other hand, phosphorous and potassium additions are known to reduce the incidence of certain pests, *e.g.*, in low phosphorous soils wireworm populations often tend to increase.

G. Water Management

Insect-pest populations can increase if irrigated plants are lusher and more attractive than surrounding plants. Irrigation of crops should be done to allow roots to get established so that the crops without any noticeable damage tolerate a certain level of white grub population (Mathur *et al.*, 2011).

H. Mulches

Natural or synthetic soil coverings may encourage or discourage insect-pests. Plastic mulches may exclude soil pests, and organic mulches may permit their control by providing a suitable habitat for their natural enemies. Crop residue mulches around fruit trees helps to control a number of pests of fruit, but the trees will need extra protection from mice, which also tend to become more abundant.

I. Crop Rotation

Crop rotation or sequence is designed to present a no host crop to insect-pests. Realistically, rotations are likely to have little effect on highly mobile foliar insects. However, less mobile foliar pests such as the Colorado potato beetle, subterranean pests or pests with one generation a year may be substantially suppressed with proper rotation. For example, to reduce insecticide applications for Colorado potato beetle by 50 per cent, potatoes have to be moved 1/4 to 1/2 mile away from previous potato crops (Weisz *et al.*, 1994).

J. Management of Alternate Hosts

Many insects reproduce on weeds or other alternate hosts and then attack the main crops. It is therefore usually desirable to destroy brambles or other weeds on uncultivated land to assist in the control of insects such as aphids, beet leafhopper and raspberry cane borer. Care must be taken, however, not to destroy nursery sites for the pest's natural enemies.

K. Trap Crops

Trap crops are used to divert insect attack away from the crop at risk by using more attractive food sources. The trap crop must usually be destroyed before the insects reproduce. This method involves the planting of a crop upon infested land so that the pest is stimulated to attack, but the knop is either removed before the pest can complete its life cycle or it will not provide all the requirements necessary for the completion of the pest's life cycle. Alternatively the trap crop may be preferentially attacked in the presence of the crop one needs to protect.

L. Pruning, Defoliation and Topping

During the dormant phase, removal and destruction of dead, diseased, or infested wood can greatly reduce overwintering stages of pest populations and thus their spread the next year, *e.g.*, mite eggs, aphids, scale and fire blight infested terminals. In apples and other fruit trees, pruning water sprouts, sucker growth or foliage that is preferred by aphids, helps control these pests. Pruning that is excessive, (*e.g.*, in relation to fertilization practices), can increase the population of certain pests such as mites, aphids, and leafhoppers.

M. Intercropping

Keeping beneficial insects in annual crops achieved by intercropping, this involves placing a crop plant and another plant within close proximity to promote insect interaction. It is one part of a comprehensive conservation plan to manipulate habitats in ways that enhance natural control. The resources provided to natural enemies include pollen, nectar, alternate prey or hosts. Many organic farms are already intercropped to maximize land use and to suppress weeds in the crop area.

2.2 Mechanical and Physical Method

A. Handpicking Method

This technique is the most practical way in certain conditions like, when cheap labour is available, insects and their eggs/egg-masses are large and conspicuous, and insects are too sluggish, have congregating behaviour and are easily accessible to the pickers. Handpicking of slow moving and visible larvae of *Pieris brassicae* L. (Cabbage butterfly) (Lepidoptera: Pieridae), lemon butterfly *Papilio demoleus* Linn. (Lepidoptera: Papilionidae), semiloopers and loopers (Lepidoptera: Noctuidae), cutworms (Lepidoptera: Noctuidae) and red pumpkin beetle *Aulacophora foveicollis* Lucas (Coleoptera: Chrysomelidae) and visible eggs/egg masses of cabbage butterfly, armyworm *Spodoptera* (Guenee) and *Mythemna* (Ochsenheimer,) spp. (Lepidoptera: Noctuidae)], and borers Pyralid borers, Noctuid borers, Crambid borers (Lepidoptera) *etc.* is an easiest, direct and excellent method of controlling them especially when their infestation is restricted to only a few plants. Collection and destruction of egg masses of top borer *Scirpophaga novella* F. (Lepidoptera: Pyralidae) in ratoon and seasonal sugarcane (*Saccharum officinarum* L.) crop reduces its endemic outbreak and losses (Saha and Dhaliwal 2012; Arif, 2019).

B. Bagging, Screening and Barriers

Bagging, screening and barriers installation is also considered very useful for protecting the crop and fruits from attack by insect pests as well as for keeping away the insect pests which either act as carrier or vectors of various fatal diseases in animals and man or create nuisance for man. For example, field bags' dragging in the maize (*Zea mays* L.) or sorghum (*Sorghum bicolor* L.) field and sugarcane (*S. officinarum* L.) ratoon crop (April/May) to collect sugarcane pyrilla *Pyrilla perpusilla* Walker (Homoptera: Lophopidae) can reduce the chances of their massive migration from maize/sorghum to sugarcane and population build-up of pyrilla at the initial growth stage of sugarcane ratoon crop. Such type of field bags can also be used for the mass collection of various grasshoppers (Orthoptera: Insecta), bugs (Hemiptera: Insecta), crickets (Orthoptera: Insecta) and other minute, small and large insects harbouring vegetation (Arif, 2019).

C Wrapping

Wrapping of individual fruits with paper bags, polythene bags, butter-paper bags or net bags protects 95 per cent of these fruits from the infestation of fruit flies. Covering whole small trees with any transparent material can reduce the attack of various insects' pests. Construction of water filled or dust (insecticide) treated drench between wheat (*Triticum aestivum* L.) field and burseem (*Trifoliumal exandrium* L.) field can reduce the migration of armyworm (*Spodoptera* spp.) larvae from wheat to berseem fields and minimize their damage on berseem. Similar type of drenches can reduce the migration of bands of locust hoppers from the breeding placed to nearest field crops (Arif, 2019).

D. Netting Method

In recent studies, special attention is given to insect exclusion netting systems in apple production. A high level of efficacy of nets was observed against Codling moth. Also, this method enables a significant reduction in pesticide use without any major risks for apple production (Alaphilippe *et al.*, 2016). PajacZivkovic *et al.* (2016), tested the effectiveness of insect exclusion netting systems in preventing the attack of Codling Moth on apple fruits. While the netting system prevents the entry of insect pests, it also serves as a barrier to beneficial insects (*e.g.*, ladybugs, true bugs and syrphid flies) which could negatively affect natural pest control services.

E. Hermetic Storage Technology (Airtight)

Hermetic simply means 'airtight'. The origin of hermetic storage dates back to antiquity. Hermetic storage (HS) technology has emerged as a significant alternative to other methods of storage that protect commodities from insects and moulds Hermetic storage is based on the principle of generating an oxygen-depleted, carbon dioxide-enriched interstitial atmosphere caused by the respiration of the living organisms in the ecological system of a sealed storage structure (Vachanth *et al.*, 2010; Obeng-Ofori, 1995; De Bruin and Murali, 2006; Donahaye *et al.*, 2001; Chauhan and Ghaffar, 2002).

F. Pheromone Traps

Pheromone trapsare being practiced successfully for the monitoring, trapping, mating disruption and management of various insect pests.

Table 20.1: Pheromone Trap Used against different Pest

Insect Common Name	*Scientific Name*	*Types of Pheromone*	*Family*	*Order*
Gypsy moth	*Lymantria dispar*	Disparlure	Lymantriidae	Lepidoptera
Cotton grey weevil	*Anthonomus grandis*	Grandlure	Curculionidae	Coleoptera
Pink bollworm	*Pectinophora gossypiella*	Gossyplure	Gelechioidea	Lepidoptera
Pine beetle	*Dendroctonus ponderosae*	Frontalinbra-vicomin	Curculionidae	Coleoptera
Oriental fruit fly	*Bactrocera dorsalis*	Methyl eugenol	Tephritidae	Diptera
European chaffer	*Rhizotrogus majalis*	Amlure	Scarabeidae	Coleoptera
Melon fruit fly	*Bactrocera cucurbıtae*	Cuelure	Tephritidae	Diptera

Arif *et al.*, 2019.

G. Light Traps

Light traps associated with toxic compound are practiced for the trapping and killing of nocturnal insect pests (Arif *et al.*, 2019).

Table 20.2: Light Trap to Attract the different Species

Common Name	Scientific Name	Different Crop	Family	Order
American Bollworm	*Helicoverpa armigera*	Mungbean, Gram, Wheat, Vegetables, Cotton, Maize	Noctuidae	Lepidoptera
Armyworm	*Spodoptera litura*	Mung bean, Gram, Wheat, Vegetables, Cotton	Noctuidae	Lepidoptera
Termites	*Microtermes* spp.	All crops, vegetables and ornamentals	Termitidae	Isoptera
Green Bug	*China viahilaris*	Mungbean, Gram, Vegetables, Cotton	Pentatomidae	Hemiptera
Grey weevil	*Myllocerus virdidanus*	Mungbean, Cotton	Curculionoidea	Coleoptera

Abbas *et al.*, 2019.

H. Pneumatic Control

In pneumatic control, insects can be dislodged from plants with negative (aspiration) or positive (blowing) air pressure, then killed by a system of turbines or collected and killed upstream in a dedicated system of the blower. With respect to the plant to be protected, distinction must be made between flexible and rigid plants. In the latter, only blowing can be used efficiently, otherwise the plant can be damaged by the machinery.

I. Shredding of Apple Leaves

Physical control methods typically affect a broad number of pest species; however, few studies have range of insects have been shown to be affected, including psyllids and rust mites (Puterka *et al*.m 2000), aphids, spider mites and leafhoppers (Glenn *et al.*, 1999); various Lepidoptera (Knight *et al.*, 2000; Unruh *et al.*, 2000) and beetles (Lapointe, 2000; Thomas *et al.*, 2004). In addition to the disruption of arthropod pests, some fungal pathogens are also killed and heat stress is reduced in trees.

J. Wash and Wax

Fruit coatings with natural waxes are known to function as modified atmosphere treatments against tephritid fruit fly immatures in fruit by reducing oxygen and raising carbon dioxide levels inside fruit (Hallman, 1997). However, when coatings are applied to fruit infested with small surface pests the mode of action is primarily physical; the organism becomes adhered and unable to move. Coatings form a major component of an integrated treatment against the mite *Brevipalpus chilensis* on cheriinoya, lime and passion fruit, from Chile to the USA (Animal and Plant Health Inspection Service, 2007).

2.3. Conservation Practices to Restore the Natural Enemies

Conservation of natural enemies involves manipulation of the environment to enhance the survival, fecundity, longevity, and behavior of natural enemies to increase their effectiveness. Such conservation efforts may be directed at mitigating harmful conditions or enhancing favorable ones. Conservation practices can be done by reducing mortality, providing supplementary resources, controlling secondary enemies, or manipulating host plant attributes to the benefit of natural enemies.

2.4. Use of Biological Control Agents

Apart from conserving native natural enemy populations which are density dependent mortality factors regulating pest populations in a given agro-ecosystem, another bio-intensive OPM option available for organic farmers is augmentative release of natural enemies that multiply during the growing season. Augmentative release may be made with either short- or long-term expectations depending on the target pest, the species of natural enemy and the crop involved and can be cost-effective. The two most commonly released bioagents are: egg parasitoids, *Trichogramma* wasps and the predatory lacewing, *Chrysoperla carnea*. *Trichogramma chilonis* is the dominant species of wasps in India. It has been used for the biological control of several pests in sugarcane, paddy, cotton and maize (Jallali and Singh, 1993; Prabhakar and Prasad, 2005)

Table 20.3: Potential Biological Control Agents for Pest Management in Organic Crop Production

Crop	*Stage*	*Pest*	*Natural Enemy Species*	*Dosage Eggs per Hectare*	*Number of Releases*
Paddy	30 days after transplanting	Stem borer	*T. japonicum*	50000	6 weekly interval
Cotton	45days after transplanting	Bollworm	*T. chilonis*	150000	6 weekly interval
Tomato	45 days after transplanting	Fruit borer	*T. brasiliensis*	50000	6 weekly interval
Suger cane	45 days after transplanting	Early shoot and stalk borer	*T. chilonis*	50000	4-6 at 10 days interval
Maize	45 days after transplanting	Stem borer	*T. chilonis*	75000	4-6 at 10 days interval
Vege-tables	30 days after transplanting	Aphids, whitefly, thrips, spodoptera	*C. carnea*	5000 grubs	3-4 at 15 days interval

2.5. Microbial Biopesticides

Microbial biopesticides have an excellent scope for management of pests in organic farming when used intelligently. Biopesticides based on Bt, NPV and GV have now been brought under the ambit of the Central Insecticide Act, 1968. Microbial bio pesticides are following types.

A. *Viral Biopesticides*

Majority of insect viruses used as biological control agents belong to the baculovirus group. These viruses are characterized by the presence of rod shaped nucleocapsid (hence "baculovirus', from baculum, meaning rod) and the ability to form proteinaceous occlusionbodies within infected cells. Baculoviruses are of two types: Nuclear polyhedrosis viruses (NPVs) and Granuloviruses (GVs) based on the type of occlusion body formed. NPVs have the following advantages for consideration in OPM initiatives: Species-specific action and hence safe to non-target organisms; nonpathogenic to most beneficial insects and hence fit admirably into vegetable crops where pesticide residues are a major problem and fit well into organic farming. Successful use of NPVs is possible in several crops (Prasad and Prabhakar, 2005).

Table 20.4: Potential Microbial Biopesticides Used for Pest Management in Organic Crop Production

Biocontrol Agents	*Effective against*	*Crops*
Bacteria		
Bacillus thuringiensis	Lepidopteran pests	Cotton, sunflower and vegetables
Entomopathogenic fungi		
Metarhizium anisopliae	Coleopteran grubs	Coconut, cotton, and
Beauveria bassiana	Lepidopteran and coleopteran	Green house vegetables
Verticilium lecanii	Aphids and whiteflies	
Entomopathogenic nematodes		
Heterorhabditis bacteriophora *Steinernema carpocapsae*	Coleopteran and Lepidopterans	Sugarcane and plantation crops
Entomopathogenic Viruses		
Nuclear polyhedrosis virus (NPV) and Granulosis virus	*Helicoverpa* and other Lepidopterans	Pulses and vegetables

Kumaranag, *et al.*, 2013.

B. *Bacillus thuringiensis (Bt)*

Bt is a naturally occurring soil bacterium that produces crystal proteins toxic to several insect species during the sporulation stage of its multiplication cycle. There are currently more than a hundred products of Bt registered for the management of important lepidopteran insect pests such as *H. armigera* (bollworm), *Plutella xylostella* (diamond back moth), *Trichoplusiani* (looper)and *A. janata* (Vimaladevi *et al.*, 1996). Bt occupies 95 per cent share of the microbial bio-pesticide market.

C. Entomopathogenic Fungi

More than 700 species are pathogenic to insects. Genera that have been most intensively investigated for mycoinsecticides include *Beauveria, Metarhizium, Verticillium, Hirsutella, Erynia, Nomuraea, Aspergillus, Aschersonia, Paecilomyces, Tolypocladium, Leptolegnia, Coelomomyces* and *Lagenidium*. The first two genera have been used on a large scale over a number of years while others have been aimed mainly at glasshouse pests or disease vectors Vimaladevi *et al.* (1996).

2.6. Plant Extracts

Indian plants represent a rich source of phytochemicals for biopesticides and raw material for agroindustries.

Table 20.5: Some Indian Plant Species Containing Phytochemicals Used in Pest Control

Plant Species	*Plant Part Used*	*Family*	*Phytochemical/ Bioactive Constituent*	*Mode of Action*
Acorus calamus	Rhizome	Araceae	Cisarmietransasarone trans-innamaldehyde propen-2, B-asarone	AF, IGR
Aesculus indica	Leaf	Hippocastanaceae	Lipophilic substances	PE
Ageratum mexicanum	Flower	Compositae	Precocene I, II	AF
Allium sativum	Rhizome	Alliaceae	Allicin	BA, FU
Angelica *glauca*	Root	Umbelliferae	Umbelliferae	FU
Annona squamosa	Seed	Annonaceae	Annonaceae	PE
Artemisia vulgaris	Leaf, Whole plant	Compositae	Capillin	FU
Azadirachta indica	Leaf seed, kernel	Meliaceae	Azadirachtin, salanin, nimbin, nimbecidin	AF, IGR, OD, PE, RE
Balanites roxburghii	Seed, Root	Balanitaceae	Diosgenin, saponin, sapogenin	AF, FU
Curcuma longa	Rhizome	Zingiberaceae	Curcumin	FU, PE, RE
Cuscuta repens	Vine	Convolvulaceae	Cuscutin	IGR
Daphne oleoides	Fruit	Thymelaeaceae	Odoracin	PE
Derris elliptica	Root	Papilionaceae	Rotenone	PE
Dioscorea sativa	Root	Dioscoreaceae	Batasin, dioscin, sapogenin	FU
Eugenia aromaticum	Flower bud	Myrtaceae	Eugenol	FU

Plant Species	Plant Part Used	Family	Phytochemical/ Bioactive Constituent	Mode of Action
Lycopersiconly copersicum	Leaf	Solanaceae	Tomatin	AF
Melia spp.	Leaf, seed	Meliaceae	Toosendanin	AF, IGR
Murraya koenigii	Leaf	Rutaceae	Carbazoloquinone	PE
Parthenium hysterophorus	Whole plant	Compositae	Parthenin	PE
Piper nigrum	Fruit	Piperaceae	Isobutylamides	PE
Plumbago zeylanica	Root	Plumbaginaceae	Plumbagin	PE
Pongamia pinnata	Seed	Papilionaceae	Pongagobol, globone,	PE
Pyrethrum cinerarifolium	Flower	Compositae	Pyrethrin I, II	PE
Santalum album	Wood	Santalaceae	Santalol	FU
Saussure alappa	Root	Compositae	Costunolide	FU
Solanum indicum	Root	Solanaceae	Solanin, solanidine	AF
Tabenaemontana divaricata	Seed	Apocynaceae	Alkaloids	OD, PE
Tagetes spp.	Root	Compositae	L-terthieny	PE
Tephrosia vogelii	Root	Papilionaceae	Amorphone/ rotenonoids	PE
Zea mays	Leaf	Gramineae	Methoxy-benzoxazolinone	AF

AF: Antifeedant; BA: Bactericidal; FU: Fungicidal; IGR: Insect growth regulator; NE: Nematicidal; OD: Oviposition deterrent; PE; RE: Repellent.

Gahukar, 2010.

Among them, neem (*Azadirachta indica*) has been in the limelight in recent years as it has been exploited for commercial products and found suitable for environmental protection. Plant products are biodegradable leaving no residues on plants; effective against insects, plant pathogens, nematodes and no insect pests; and cheaper, non-polluting, easy to prepare and compatible with several biopesticides. They are therefore recommended for large scale application in organic farming.

CONCLUSION

Pest management in organic farming is achieved by using appropriate cropping techniques, biological control, and natural pesticides (mainly extracted from plant or animal origins). Weed control, the main problem for organic growers, can be managed through cultural practices including mechanic cultivation, mulching, and flaming. Organic farming is characterized by higher diversity of arthropod fauna and conservation of natural enemies than conventional agriculture. Pest management in organic farming depends mainly on crop husbandry and biological control. The prohibition of synthetic fertilizers and pesticides leads to conservation of natural enemies including predators and parasitoids. The absence of harmful pesticides also increases diversity of pollinators of crops and minimizes pesticide residues in food products. The community of microorganisms flourishes well in organically managed farms leading to increased organic matter decomposition, soil fertility and sustainability of the ecosystem

REFERENCES

Abbas M, Ramzan M, Hussain N, Ghaffar A, Hussain K, Abbas S and Raza A.2019. Role of Light Traps in Attracting, Killing and Biodiversity Studies of Insect Pests in Thal. *Pakistan Journal of Agricultural Research.* 32(4): 684.

Alaphilippe A, Capowiez Y, Severac G, Simon S, Saudreau M, Caruso S and Vergnani S. 2016. Codling moth exclusion netting: An overview of French and Italian experiences. IOBC-WPRS Bull, 112, 31-35 and damage in potato. *Biol. Control.* 4: 163-69.

Animal and Plant Health Inspection Service. 2007. Treatment Manual. Washington, DC and Riverdale, MD: US Department of Agriculture. Available at www.aphis.usda.gov/intport_export/plants/manuals/ports/downloads! treatment.pdf.

Arif MJ, Gogi MD, Sufyan M, Nawaz A, Sarfraz. 2019. Principles of Insect Pests Management Department of Entomology, University of British Columbia, Canada.

Chauhan YS, Ghaffar MA. 2002. Solar heating of seeds a low cost method to control bruchid (*Callosobruchus* spp.) attack during storage of pigeon pea. *J Stored Prod. Res* 2002;38: 87-91.

De Bruin T and Murali V. 2006. Quality Preservation of Stored Cocoa Beans in the Tropics.(Unpublished).GrainPro Document Number SL2299TB0905. GrainPro, Inc. Concord, MA USA.

Devi VPS, Prasad YG. and Rajeswari B. 1996. Effect of *Bacillus thuringiensis* and neem on castor defoliators – *Achaea janata* (Linnaeus) and *Spodoptera litura* (Fabricius).*Journal of Biological Control.* 10: 67-71.

Donahaye EJ, Navarro S, Sabio G, Rindner M, Azrieli A and Dias R. 2001. Reflective covers to prevent condensation in sealed storages in the tropics. *Executive Printing Services, Clovis,* CA, U.S.A, 227-230.

Elshafie H. 2019. Insect Pest Management in Organic Farming System. Multifunctionality and Impacts of Organic Agriculture. DOI: http: / /dx.doi.org/10.5772/intechopen.84483

Gahukar RT. 2010. Role and perspective of phytochemicals in pest management in India. Current Science.98(7).

Glenn DM, Puterka GJ, Vanderzwet T, Byers RE and Feidhake C. 1999. Hydrophobic particle films: a new paradigm for suppression of arthropod pests and plant diseases. Journal of Economic Entomology. 92: 759-771.

Gomiero T, Pimentel D and Paoletti MG. 2011. Environmental impact of different agricultural management practices: Conventional vs. organic agriculture. Critical Reviews in Plant Sciences.30: 95-124

Haldhar SM, Jat GC, Deshwal HL, Gora JS and Singh D. 2017.Insect Pest and Disease Management in Organic Farming.*Towards Organic Agriculture*: 359-390.

Hallman GJ. 2001. Irradiation as a quarantine treatment. In Food Irradiation: Principles and Applications, ed. R Molins, New York. Wiley, 113-30.

Jallali SK and Singh SP. 1993.Superior strain selection of the egg parasitoid Trichogrammachilonis Ishii.Biological parameters. Journal of Biological Control.7: 57-60.

Knight AL, Unruh TR, Christianson BA, Puterka GJ and Glenn DM. 2000. Effects of a kaolin-based particle film on oblique banded leaf roller (Lepidoptera: Tortricidac). Journal of Economic Entomology.93: 744749.

Kumaranag KM, Kedar SC and Patil MD. 2013.Insect Pest Management in Organic Farming Systems. *Popular Kheti* 1(4): 3.

Lapointe SL. 2000. Particle film deters oviposition by *Dioprepes abbreviatus* (Coleoptera: Curculionidae). *Journal of Economic Entomology*. 93: 1459-1463.

Mathur YS, Bhatnagar A and Singh S. 2011. Phytophagouswhitegrubs of Indian Bioecology and Management. AINP on whitegrubs and other soil Arthropods Department of Entomology, ARS, Durgapura, Swami Keshvanand Rajasthan Agricultural University Pp.28.

Matyjaszczyk E. 2015. Prevention methods for pest control and their use in Poland.*Pest Management Science*. 71: 485-491.

Mohan S, Devasenapathy P, Vennila C and Gill M S. 2013. IPM Booklet for Pest and Disease Management: *Organic Ecosystem*. P. 21-22.

Nicholas JM. 2005.Biological and cultural controls. Non-pesticide alternatives can suppress crop pests. *California Agriculture*. 59: 23-28.

Nissen RJ, George AP and Topp BL. 2005b.Producing super Sweet and firm peaches and nectarines. *ActaHorticulturae*. 694: 311-314.

Nissen RJ, George AP, Waite G, Lloyd A and Hamacek E. 2005a. Innovative new production systems for lowchillstonefruit in Australia and South-East Asia: a review. *Horticulturae*. 694: 247-251.

Obeng OD. 1995. Plant oils as grain protectant against the infestation of *Cryptolestes pusillus* (Schorr.) and *Rhyzopertham dominica* (Fab.) in stored grain. *Entomologia Experimentaliset Applicata*. 77: 133-139.

PajacZl, Jemric TFM and Buhin J, Baric B. 2016. Influence of different netting structures on codling moth and apple fruit damages in northwest Croatia. *Agric. Conspec. Science*. 81: 99-102.

Patil A P, Nawale RN, Azri DS and Moholkar P. 1988.Field screening of onion cultivars for their reaction to thrips. *Indian Cocoa Arecanut and Spices Journal*. 12: 10-11.

Peterson G, Allen CR, Holling CS. 1998. Ecological resilience, biodiversity, and scale.*Ecosystems*. 1: 6-18.

Prabhakar M and Prasad YG. 2005. Trichogramma egg parasitoid for biological control of insect pests. Technical brochure 2/05, CRIDA, Hyderabad.

Puterka GJ, Glenn DM, Sekutowski DG, Unruh TR and Jones SK. 2000.Progress toward liquid formulations of particle films for insect and disease control in pear.*Environmental Entomology*. 29: 329-339.

Saha LR and Dhaliwal GS. 2012. Handbook of Plant Protection, 2nd Edition, Kalyani Publishers, New Delhi, India.

Stern VM, Smith RF, van den Bosch R, Hagen KS. 1959. The integrated control concept. Hilgardia. 29: 81-101.

Thakur K, Sharma A and Sharma K. 2021.Management of agricultural insect pests with physical control methods.*The Pharma Innovation Journal*. 10(6): 306-314.

Thomas AL, Muller MF, Dodson BR, Ellersieck MR and Kaps M. 2004. A kaolin-based particle film suppresses certain insect and fungal pests while reducing heat stress in apples. *Journal of the American Pomology Society*. 58: 42-51.

Unruh TR, Knight AL, Upton J, Glenn DM and Puterka GJ. 2000. Particle films for suppression of the codling moth (Lepidoptera: Tortricidae) in apple and pear orchards. *Journal of Economic Entomology*. 93: 737-743.

Vachanth MC, SubbuRKM, Preethi R and Loganathan M. 2010. Controlled Atmosphere Storage Technique for Safe Storage of Processed Little Millet. Academic.*Journal of Entomology*. 3(1): 12-14.

Weisz R, Smilowitz Z and Christ B. 1994. Distance, rotation, and border crops affect Colorado potato beetle (Coleoptera: Chrysomelidae) colonization and population density and early blight (*Alternaria solani*) severity in rotated potato fields. *Journal of Economic Entomology*.87(3): 723-729.

Westerman PS, Wes JS, Kropff MJ, van der Werf W. 2003.Annual losses of weed seeds due to predation in organic cereal fields.Journal of Applied Ecology.40: 824-836.

Chapter 21

Determination of Biological Activity of Soil by Dehydrogenase Activity

Pratibha Kumari[1], Babloo Sharma[2] and Reena Kumari[3]

[1]Department of Soil Science and Agricultural Chemistry, Institute of Agricultural Science, Banaras Hindu University, Varanasi–221 005, Uttar Pradesh
[2]Krishi Vigyan Kendra (Swami Keshwanand Rajasthan Agricultural University), Pokaran, Jaiselmer–345021, Rajasthan
[3]Agricultural Engineering Department, N.M. College of Agriculture, Navsari Agricultural University, Navsari–396450, Gujarat

Principle

Biological activity or index of a soil is the function of number of organisms present in soil and their physiological efficiency. The rate of respiration can be used as an index of the biological activity of soil as it reflects the physiological efficiency of the organisms. Among all enzymes in the soil environment, dehydrogenases are one of the most important, and are used as an indicator of overall soil microbial activity, because they occur intracellular in all living microbial cells. Monitoring of dehydrogenases, which are respiratory enzymes and integral part of all soil organisms, give a measure of biological activity of soil at a given time.

During respiration, biological oxidation of reduced compounds occurs which is catalyzed by dehydrogenases. During this process energy is evolved. The process can be represented as:

$$RH_2 + A \rightarrow R + AH_2\ (2H^+.2e^-)$$

Where, RH_2 represents a reduced compound (hydrogen donor) and A is the electron acceptor, which in aerobic organisms is oxygen. Under anaerobic condition compounds like 2,3, 5-triphenyl tetrazolium chloride (TTC) can act as an electron acceptor. In the process TTC gets reduced to a pink coloured compound triphenyl formazan (TPF) which can be quantitatively extracted by methanol and measured colorimetrically.

$$TTC + 2H^+.2e^- \rightarrow TPF$$

The dehydrogenases can be assayed and activity can be expressed as rate of formation of TPF from TTC. Higher the biological activity faster will be the formation of TPF.

Material Required

- ☆ Air tight screw cap
- ☆ test tube
- ☆ Spectrophotometer
- ☆ Incubator
- ☆ Whatman No.1 filter paper

Reagents

1. 3 per cent 2,3,5- Triphenyl tetrazolium chloride (TTC): Dissolve 3 g TTC in 100 ml of distilled water. Store in an amber coloured bottle.
2. Methanol (AR grade)
3. 1 per cent Glucose solution: Dissolve 1 g of glucose in 100 ml of distilled water.
4. Standard Triphenyl formazan (TPF, 100 ppm): Dissolve 100 mg of TPF in 100 ml of distilled water.

Procedure

1. Put 1 g of representative air-dried soil in air tight screw capped test tube of 15 ml capacity.
2. Moist the soil to its water holding capacity (40-50 per cent).
3. Add 0.2 ml of 3 per cent TTC solution in each of the tubes to saturate the soil.
4. Add 0.5 ml of 1 per cent glucose solution in each tube. Gently tap the bottom of the tube to drive out all trapped oxygen. A water seal is formed above the soil. Ensure that no air bubbles are formed.
5. Incubate the tubes at 28± 0.5°C for 24 hours.
6. After incubation add 10 ml of methanol and shake vigorously and allow to stand for 6 hours.

7. Withdraw clear pink coloured supernatant liquid and read the absorbance at a wavelength of 485 nm (blue filer).
8. Extrapolate TPF formed from the standard curve drawn in the range of 10 to 90 mg TPF mL^{-1}.
9. Express the results as mg TPF formed per hour per g soil.

Determination of Soil Microbial Biomass Carbon (SMBC)

Principle

In the fumigation-extraction method, a direct measurement of C and other nutrients contained in microbial biomass is carried out. Overnight fumigation of chloroform is done to kill all the organisms in soil samples. The microbial biomass constituents released by $CHCl_3$ fumigation treatment can be extracted directly through chemical extractants. The readily oxidizable C contained in the extractant can be measured through standard chemical procedures.

Material Required

- ☆ Moisture box Separating funnel
- ☆ Glass beads, conical Flask (500 mL), beaker (50 mL), funnels other glassware Vacuum desiccator and Vacuum pump
- ☆ Rotary shaker Hot plate
- ☆ Whatman No.1 filter paper

Reagents

1. Distilled chloroform ($CHCl_3$)
2. Conc. H_2SO_4
3. 0.5 K_2SO_4: Dissolve 87.126g of K_2SO_4 in distilled water and final volume made up to 1Lt.
4. Orthophosphoric acid
5. 0.2*N* $K_2Cr_2O_7$: Dissolve 0.9808g $K_2Cr_2O_7$ in distilled water and final volume made up to 100mL.
6. 0.05*N* Ferrous Ammonium Sulphate (FAS): Dissolve 19.6g FAS in distilled water and add 12 drops of H_2SO_4 and final volume made up to 1Lt.
7. Ferroin/Diphenylamine indicator

Procedure

1. Weight 50g of moist field soil sample each in 3 sets, one for complete killing the microbes and another for keeping in refrigerator to predict the normal death rate, last one for determination of soil moisture content gravimetrically.

Purification of chloroform - Take 50 mL chloroform and just half of it *i.e.*, 25 mL H_2SO_4 and it should mix vigorously, after transferring in a separating funnel, acid should be discarded *i.e.* on the bottom part. This step should perform 3 times and finally, 3 times washing with double volume distilled water should do and in those steps bottom part should collected. This purification is very much important if it not purified properly then chloroform would not boil.

2. All the soil samples and pured 50 mL chloroform should keep in vacuum desiccator. Inner surface of the desiccator should line with moisten filter paper, after switch on the pump it would keep on until boiling of chloroform then it would kept for 5minutes. To reduce the excess bumping of $CHCl_3$, glass beads are add into it. Then after switch off the pump it would kept in dark for 24hrs. Next day release the vacuum and take out the beaker containing chloroform, and the inner paper lining. Perform back suction for five to six times to ensure removal of any excess/adhere $CHCl_3$ vapor.
3. Take out the non fumigated soil sample from the refrigerator, and thawing it immediately.
4. Take both the fumigated and non fumigated soil samples in 500 mL conical Flasks each and 100 mL 0.5M K_2SO_4 (1:5:: soil: extractant) should add in each sample. Shake for half an hour, then filtration done with Whatman No.1 filter paper.
5. Transfer 10mL of the filtrate in 500mL conical Flask. Add 2 mL $K_2Cr_2O_7$ (0.2 *N*), 10 mL conc. H_2SO_4 and 5 mL of orthophosphoric acid to each Flask. Run at least 2 blanks with 10 mL distilled water each, along with the acids mentioned above.
6. Put funnels in every Flask (for Refluxing) and keep those on hot plate at 100°C for ½ hour.
7. Take out the Flaks and add about 250 mL of distilled water immediately. Allow the contents to cool to room temperature.
8. Add 2-3 drops of Ferroin indicator and titrate against 0.05*N* FAS to get a brick-red end point.

Calculations

1. Soil water content (WS):

$$\text{WS per cent} = \frac{[\text{Wt. of wet soil (g)} \times \text{wt. of oven dry soil (g)}]}{\text{Wt. of oven dry soil (g)}} \times 100$$

2. Weight of soil sample (oven dry weight equivalent) taken for microbial biomass measurement (MS):

$$MS\ (g) = \frac{\text{Wt. of wet soil (g)}}{\{100+ WS(\text{per cent})\}} \times 100$$

3. Total volume of solution in the extracted soil (VS):
   VS (mL) = soil water + extractant volume
   = [wt. of wet soil(g)-wt. of oven dry soil(g)]+extractant volume (mL)

4. Determination of extractable carbon (EC in micro g mL^{-1}):
   a) Standardization of FAS solution:
      Normality of FAS (XN)
      $$= \frac{\text{Volume of } K_2Cr_2O_7 \text{ (2ml) x Strength of } K_2Cr_2O_7 \text{ (0.2N)}}{\text{Average titre value for the blank (ml)}}$$
   b) Determination of volume of $K_2Cr_2O_7$ solution consumed by FAS in any sample (Y) ml Y
      $$= \frac{\text{Normality of FAS (XN) x Titre value (mL)}}{\text{Normality of } K_2Cr_2O_7 \text{ (0.2N)}}$$
   c) Vol. of $K_2Cr_2O_7$ consumed for oxidizing easily mineralizable C in 10 ml of extractant = 2-Y mL
   d) Extractable C (EC in µg mL^{-1}):
      1 ml of 1N $K_2Cr_2O_7$ oxidizes 0.003 g of C
      1 ml of 0.2 N $K_2Cr_2O_7$ oxidizes 0.0006 g of C *i.e.* 600 µg of C (2-Y) ml of 0.2N $K_2Cr_2O_7$ oxidizes 600*(2-Y) µg of C
      So, the amount of extractable C (EC) EC(µgmL^{-1})
      $$= \frac{600 \times (2-Y)}{10}$$

5. Total weight of extractable C in the fumigated (EC_F) and non-fumigated (EC_{NF}) soil sample:

$$EC_F \text{ or } EC_{NF}\ (\mu g\ g^{-1} \text{ soil}) = \frac{EC\ (\mu g\ mL^{-1}) \times VS\ (ml)}{MS\ (g)}$$

6. Soil microbial biomass carbon in soil (SMBC)

$$= \frac{[EC_F - EC_{NF}]}{KEC}$$

Where, $K_{EC} = 0.25 \pm 0.05$ and represents the efficiency of extraction of microbial biomass carbon.

Precaution

1. Chloroform is a well known carcinogenic compound, it also gets readily vaporized. Musk and hand gloves are necessary during handling this chemical. If it not properly pure then it will not vaporized.
2. We should add Glass beads to reduce bumping of $CHCl_3$ during boiling.
3. Back suction should done to reduce erroneous result as $CHCl_3$ also contain carbon.

Monitoring Organic Matter Decomposition in Soil through Carbon Dioxide Evolution

Principle

The method involves absorption of CO_2 evolved during a given period of time in a known volume and strength of alkali (*e.g.*, NaOH or KOH). When CO_2 is absorbed in NaOH it is converted to Na_2CO_3. The excess of NaOH is titrated against standard HCL. Prior to titration with HCl, few drops of saturated $BaCl_2$ solution is added to NaOH solution to precipitate the Na_2CO_3 as $BaCO_3$, otherwise CO_3^{2-} in Na_2CO_3 will consume HCL and underestimate the CO_2 evolved.

$$2NaOH + CO_2 \rightarrow Na_2CO_3 + CO_2$$

$$NaOH + HCl \rightarrow 2NaCl + H_2O$$

$$Na_2CO_3 + Bacl_2 \rightarrow BaCO_3 + 2NaCl$$

$$Na_2CO_3 + HCl \rightarrow 2NaCl + H_2CO_3$$

The volume of NaOH that absorbs CO_2 is taken for calculation of amount of CO_2 evolved.

Material Required

Cylindrical glass jar with lid (500 ml) Glass scintillation vial (10 ml)

Glass beaker (100 ml) Rubber stoppers Burettes and pipettes Asbestos sheets

Reagents

1. Standard NaOH solution (0.5N): Dissolve 20 g of NaOH in distilled water and make volume to 1 L.
2. Standard HCl solution (0.5N): Dilute 1.4 ml conc. HCl to 1L.
3. Saturated solution of $Bacl_2$

4. Phenolphthalein indicator: Dissolve 0.4 g phenolphthalein in 1:1 H_2O : ethanol mixture and make up the volume to 100 mL.

Procedure

1. Conical flask of 500 ml capacity are filled with 50 or 100 g of soil.
2. Moist the soil to its water holding capacity (40-50).
3. A test tube containing 10 ml of N/10 NaOH solution is hung in each flask with the help of thread.
4. The flask are closed with rubber stopper and sealed with molten wax and incubated at 30°C ±1 or 2°C.
5. The CO_2 evolved is absorbed by alkali in the test tube and measured titrimetrically at desired intervals.
6. During each estimation the contents of the test tube is transferred to a flask followed by addition of 5 ml of 3 N $BaCl_2$ solution to stop further CO_2 absorption.
7. The residual amount of NaOH in the flask is measured by titrating against N/10 HCl using 2-3 drops of Phenolphthalein indicator.
8. Note the reading where pink colour of solution changes to colourless.

Calculation

mg CO_2 = (B-V) × (N×E)

where,

B = Volume (ml) of standard acid needed to titrate the trap solution from the empty flask (blank) to the end point

V = Volume of HCl required to titrate sample to end points N = Normality of the acid in milli equivalents mL^{-1}

E = The equivalent weight of C in CO_2; E= 22 if the data is to be expressed as CO_2 (*i.e.* mg CO_2)

REFERENCES

Chhonkar. P. K., Bhadraray S., Patra. A. K. and Purakayastha. T. J. (2015) "Determination of soil biological activity by monitoring dehydrogenase activity. Westville publishing house, New Delhi.pp.57-59.

Chhonkar. P. K., Bhadraray S., Patra. A. K. and Purakayastha. T. J. (2015) "Determination of soil microbial biomass carbon by fumigation-extraction method. Experiments in soil biology and biochemistry".Westville publishing house, New Delhi.pp.72-78.

Chhonkar. P. K., Bhadraray S., Patra. A. K. and Purakayastha. T. J. (2015) "Monitoring organic matter decomposition in soil through Carbon Dioxide evolution: Alkali Trap Method. Westville publishing house, New Delhi. pp.67-71.

Chapter 22

Organic Animal Husbandry

Sachin Kumar Shandilya

Assistant Professor (AH and Dairying), CSSS (PG) College, Machhra, Meerut, Uttar Pradesh

ABSTRACT

The animal production system may be broadly classified as traditional, conventional and organic with their distinct characteristics. While traditional and conventional production systems are well established and their respective features known widely, the organic animal husbandry system has emerged only recently, which is still evolving. Many believe that the organic animal husbandry is same as traditional animal husbandry practiced in India since centuries, which, of course, it is not. Organic animal production is a means of food production with a large number of rules directed towards a high status of animal welfare, care for the environment, restricted use of medicinal drugs and the production of a healthy product without residues (pesticides or medicinal drugs). Organic dairy farming aims at sustaining animal health and production systems in the existing agro-climatic conditions of the country. The main emphasis is to create resources for the production of organic high quality milk and highly processed organic dairy products with a well-described history and a composition that differs from inorganic production system. In this chapter emphasis has been given on organic dairy farming status, advantages and constrain in Indian conditions.

Keywords: *Animal husbandry, Organic animal husbandry, Organic dairy farming, Status.*

Various Aspects of Organic Dairy Farming

Care for Animal

Organic milk is produced with better care and kindness to animals in comfortable conditions. The animal are reared and maintained usually in open system of grazing in pastures which are not treated with chemical fertilizers and pesticides. Animals are free to express their natural/normal behavior. All these are environment friendly for the animals which provide better animal welfare.

Farm Land Wildlife

There are reports that more birds, butterflies, beetles, bats and wild flowers are seen on organic farms than conventional farms. The hedges on organic farms help to create 'wildlife corridors' providing cover for small mammals such as rabbits, hares, birds and other species as they travel from place to place.

Healthy Soil

In non-organic farming a limited number of crops are grown intensively and soil is quickly stripped of many essential nutrients unless synthetic chemical fertilizers are applied. Over time the soil is depleted of vital organic materials and nutrients and can take many years to recover. In contrast, organic farming maintains a healthy and balanced soil which is rich in natural nutrients. It provides an eco friendly-system for birds, insects and earthworms.

Watercourses and the Nitrates

The fast growing short span crops of non-organic farming rely on the use of synthetic nitrates to enhance nitrate content of soil. Usually about one fifth of these nitrates are washed away with rainwater and into the watercourses. Excessive nitrates over-stimulate the growth of algae and poison the water and, in turn, the fish and the animals that feed on them. The use of natural farmyard manure and planting legume crops to fix nitrogen, organic farmers ensure that nitrates are released slowly, naturally and not in excess.

Environment

The use of farm manures, excreta and natural methods of disease and pest control inorganic farming provide considerable environmental benefits over non-organic farming. Milk of bovines and other species is also an ideal environmental indicator to know level of pollutants and pesticides in the environment. The feed that goes into the mouth of a cow gets processed into milk. Among the global organic industry's top producers, organic dairy and poultry production have shown stronger growth rates than organic beef and pork production. Some reasons for this trend may include varying national and private requirements for transitioning from conventional to organic livestock production, competition for organic grain between consumers and livestock, and a slower evolution of organic production standards for beef and pork than for organic production.

Principles of Organic Dairy Farming

The main goal of organic dairy farming is food quality and safety, consumer protection, animal health and welfare and the sustainability. General principles of organic dairy farming include livestock management, conversion period, and introduction of animals from outside, breeds and breeding, mutilations, feeding, health coverage, transport and slaughter and cleaning and disinfection of sheds.

Livestock Management

Management techniques should take care of physiological and ethological needs of the animals. The livestock management of the animals should take into account the behavioural needs of the animals *viz.*; 1) provided with sufficient space for free movement, sufficient fresh air and natural daylight, protection against excessive sunlight, temperature, rain and wind, enough lying and/or resting area, and access to fresh water and feed; 2) construction materials which might detrimentally affect human or animal health should not be used; 3) all animals shall have access to open air and/or grazing appropriate to the type of animal and the season except in cases where the specific farm or settlement structure prevents such access and areas where feeding of animals with cut and carry fodder is more sustainable way to use land resources than grazing and 4) herd animals shall not be kept individually however, male animals, small holdings, sick animals and those about to give birth are exceptions.

Conversion Period

The whole farm including livestock should be converted according to the set standards *viz.*; animal products may be sold as "product of organic agriculture" only after the farm or relevant part of it has been under conversion for at least twelve months. The length of conversion period may vary. With regards to dairy production this period shall not be less than 30 days. For organic meat, the conversion period should have been 12 months.

Introduction of Animals from Outside

All animals should be born and raised on the organic holding. When trading or exchanging livestock, it should preferably take place between organic farms or as part of a long-term cooperation between specific farms. When organic animals are not available, the calves (limiting to 4 weeks old) which have received colostrum and are fed a diet consisting mainly of full milk are allowed. Breeding stock from conventional farms limiting to 10 per cent of the adult animals each year can be introduced. More than 10 per cent introduction of animals from outside is allowed in cases like unforeseen severe natural or man-made events.

Breeds and Breeding

Breeds adapted to local conditions should be chosen. Breeding techniques should be natural. Breeds used for breeding should be compatible which can copulate and give normal birth. Artificial insemination and Embryo transfer techniques are not allowed. Use of hormones and induced birth are not allowed unless applied to individual animals on health grounds. The use of genetically engineered animals is not allowed.

Mutilations

Mutilations are not allowed. Castration, tail docking of lambs, dehorning, ringing *etc.* are however allowed for animal welfare if these are unavoidable like in case of irreparable injury, disease and malignancy. Suffering shall be minimized and anaesthetics where appropriate may be used.

Feeding

The livestock should be fed 100 per cent organically grown feed of good quality. The feed to the animals should be offered in a form which allows them to execute their natural feeding behaviour and digestive needs. The diet should be balanced as per the nutritional needs of the animals. The products from the organic food processing industry can be used. The antibiotic, coccidiostats, medicinal substances or any other substances intended to stimulate growth or production should not be used in animal feed. The other products like synthetic growth promoters or stimulants, synthetic appetizers, preservatives (except when used as a processing aid), artificial colouring agents, urea, farm animal by-products (*e.g.* abattoir waste), droppings, dung or other manure (all types of excreta), even if technologically processed, feed (soya and rape seed meal) subjected to solvent (*e.g.* hexane) extraction or the addition of other chemical agents, and genetically engineered organisms or products should not be added to the animal feed. Vitamins, trace elements and supplements of natural origin may be used. The animals shall have daily access to roughage. The fodder preservatives (bacteria, fungi and enzymes), by-products of food industry (*e.g.* molasses), and plant based products are allowed. Each lactating animal should be given minimum weaning time. Young stock shall be raised on organic milk, preferably from their own species. In emergencies, the use of milk from non-organic farming systems or dairy based milk substitutes so long as they do not contain antibiotics or synthetic additives should be allowed.

Health Coverage

Prevention of diseases is the golden rule in organic agriculture. Management practices which can boost resistance against diseases and can prevent infections should be followed. When illness does occur, the aim should be to find the cause and prevent future outbreaks by changing management practices. Use of natural medicines and methods, including homeopathy, ayurvedic, and unani shall be emphasized. Sick and injured animals shall be given prompt and adequate treatment. Conventional veterinary medicines may be used in situations when no other justifiable alternative is available. The withholding period for using animal products shall be at least double the legal period which is specified for each medicine. Use of synthetic growth promoters, hormones for heat induction and synchronization unless used in an animal for treating reproductive disorders, is prohibited. Vaccination shall be used only when diseases are known or expected to be a problem in the viscinity of the farm and where these diseases cannot be controlled by other management techniques. Only legally required vaccinations are allowed. Genetically engineered vaccines are prohibited.

Transport and Slaughter

The animals should have minimum stress and inspected regularly during transport. They should be watered and fed as per weather conditions and duration of the transport. Each animal shall be stunned before being bled to death. The equipment used for stunning should be in good working order. Exceptions can be made according to cultural practice. Where animals are bled without prior stunning, the slaughter should take place in a calm environment. Handling during transport and slaughter shall be calm and gentle. The use of electric sticks and such instruments are prohibited. The standards set for transport and slaughter in terms of stress caused to the animal and person in charge, fitness of the animal, loading and unloading, mixing different groups of animals, animals of different sex, quality and suitability of mode for transport and handling equipment, temperature and relative humidity, hunger and thirst *etc.* should be considered. No synthetic, tranquilliser or stimulant shall be given prior to or during transport. Each animal or group of animals shall be identifiable during all steps of transport and slaughter. The journey time to the slaughterhouse shall not exceed eight hours.

Cleaning and Disinfection of Sheds

Use of agents *viz.*; potassium and sodium soap, water and steam, lime, sodium hypochlorite (*e.g.* as liquid bleach), caustic potash, hydrogen peroxide, natural essences of plants, citric, formic, lactic, oxalic and acetic acid, nitric acid, phosphoric acid, alcohol, formaldehyde and sodium carbonate is authorized.

Advantages and Limitations of Organic Livestock Production

Organic food products are generally considered to be of better quality due to strict compliance of quality enhancing measures. Yet, the organic livestock products are not completely free from criticism in quality front. Some of the reported benefits as also the limitations of organic livestock production are:

Advantages of Organic Livestock/Milk Production

1. It ensures strict animal welfare measures
2. Recognize animal comfort and animal behaviour

3. It is better for environment;
4. Better sustainability of production;
5. Continuation of traditions and culture;
6. Boost to traditional technologies;
7. It is 'safer' (antibiotic/chemical/drug/pesticide free);
8. It has better protein quality (proteins of organic pork are less denatured than those of traditional pork during cooking);
9. It has lower content of carcinogenic nitrogen, thus, reduces carcinogenic problems;
10. Organic milk and meat taste equal to or better than traditional/ conventional milk and meat;
11. Veterinary costs are generally significantly lower on organic farms than conventional farm;
12. Less importance on high live weight gains facilitate crossing breeds with higher intramuscular fat in meat for better relish when eaten;
13. Greater motion reduced abdominal fat and favoured muscle mass development in broilers;
14. Motor activity increased breast percentage in chicken;
15. Motor activity favours myogenesis against lipogenesis;
16. Lower pH in organic meat due to better welfare and reduced pre-slaughter stress and thus, reduced consumption of glycogen and
17. No drug or antibiotic residue in milk and meat.

Alongside the advantages, there are certain limiting implications associated with a shift to organic livestock production, *viz.*

Limitations of Organic Livestock/Milk Production

Though our farmers are resources poor, they cannot afford to remain aloof to the changes in the global market and requirements for exporting the animal husbandry products in the liberalized economy. However, following are important limitations in adopting organic livestock production:

1. Indian agriculture has been moving from low inputs usage to increasing use of high yielding varieties, fertilizers, pesticides, crossbreds, allopathic medicines, antibiotics, growth promoters, feed additives *etc.* organic milk and meat production may further decrease the availability of milk and meat;
2. The farm area required per animal under organic farming is higher than the conventional production system, due to higher green fodder requirement and the large locomotion area;

3. As artificial insemination (A.I) is not allowed under organic milk production, the massive crossbreeding programme promoted by our government and co-operatives would have its adverse implications.
4. Prolonged withdrawal period required after any veterinary treatment with drugs, make farmers reluctant to call a veterinary surgeon, reaching to a delay in the treatment of sick animals and a deterioration in animal welfare;
5. The yield per animal in organic farming is comparatively lower than conventional farming and the cost of production of organic milk/meat is also very high.
6. The inadequate availability of ayurvedic, homeopathic and other herbal products will hamper the organic production.
7. The organic certification process will be too expensive for our average dairy farmers.
8. As majority of agriculture and dairy farmers are illiterate and less informed/aware, they do not usually keep written farm records.
9. Lack of training opportunities and incentives for organic milk production and market development.
10. Non availability of nearby markets coupled with unreliable transport facilities will directly affect the price of the produce.
11. Processing or preservation of meat and meat products is difficult without use of certain chemicals *e.g.* trisodium polyphosphate, sodium nitrate; blanket rejection of preservatives may have serious effect on food supply and safety of foods;
12. Low involvement of government agencies, lack of subsidies, lack of clear cut policy, less effort to develop domestic market and lack of proper marketing network of the organic produce- are some of the constraints in organic products marketing.

Owing to these limitations, organic farming is often criticized and questioned in developing country context, where there is already a shortage of foods.

Favourable features for organic animal husbandry development in India or the opportunities for India:

The threats apart, there are strong reasons for India to focus attention on organic livestock production. Some of the encouraging factors are:

1. Unlike in western developed countries, where, specialized farming is common, farms in India are mostly diversified in term of crops grown, species and breeds of livestock raised (Crop- livestock integrated/mixed farming system). On-farm diversity is considered good for the organic farming. The requirements of fodder and feed are met on-farm.

The dependence on market for external inputs is reduced to minimum. Under organic production management externally purchased inputs are discouraged. Indian traditional animal husbandry system has potential for conversion to organic animal husbandry system.

2. Largest livestock, poultry population and number one milk producer in the world, some proportion of this vast sector may be developed as organic in the areas, where, it is comparatively easy to convert to organic farming.
3. In India native breeds are well adapted to local situations. Being hardy, resistant to diseases, health and maintenance cost is very low.
4. Indigenous Technical Knowledge (ITK) available in India may provide effective substitute to veterinary care;
5. The use of agro-chemicals is almost nil in some parts of India. ideal for the development of organic livestock production;
6. Indian native livestock are less susceptible to diseases and stress, need for allopathic medicines/antibiotics is very less;
7. Grass based extensive production systems/forest based animal production prevalent in some parts of India have potential for conversion into organic animal husbandry;
8. Literacy is on the rise, media is making the consumers aware and concerned about animal welfare issues and health foods- it may boost domestic consumption of organic foods; and
9. Growing domestic market for organic products in India may help boost organic market at country, and regional level,
10. The GOI is strongly committed to animal disease control and eradication. Rinderpest (RP) is not reported in India since 1995, OIE has declared it free from RP. Foot and Mouth 'Disease (FMD) control programme is being implemented aggressively. India never had BSE and many infectious diseases are not reported in India in last 10 years. The more controlled animal health environment is needed for the organic animal husbandry. The Indian native livestock breeds are less susceptible to diseases and stress, thus, ideally suited for the organic animal production.
11. Existing extensive livestock production systems are easier for conversion to organic production. The areas with low to negligible external input use like remote hilly terrains, tribal areas and certain north eastern states have favourable conditions for the development of organic farming.
12. Small animals like goat, sheep, pigs, and rabbits may make good beginning in organic meat and leather production. Also, domestic market for organic milk and milk products *viz. ghee* can be developed

with ease, Organic leather could be a valuable export commodity from India. Backyard poultry systems with indigenous breeds have potential for organic poultry production. Some consumers prefer milk, meat and eggs from native breeds for which they pay even premium prices readily,

Strategies for Promoting Organic Livestock Milk Production

1. Production standards defining the practices of organic livestock/milk production has to be formulated, legislated, and implemented through the help of Govt. Agencies.
2. A duly constituted certification body of authority must certify the organic production and processing of milk without any fees and charge practices.
3. The retailing of organic dairy products can be achieved through general stores, supermarkets and farms sales.
4. Since there is a greater awareness on healthy issues among high income professionals and educated consumers "a special market"/or big bazar could be created in towns and cities to promote organic food sale.
5. Dairy farmers must be imparted training in organic dairy farming.
6. Studies should be conducted on economics of organic milk production.
7. Awareness must be among the consumers on advantages of organic milk and promoting Indian organic products at international fairs and media.
8. Favourable prices and subsidies for organic milk production will attract large number of producers.
9. Organic farming system and production should be given adequate focus in recently launched Kisaan Call Centers, television channel for farmers and new insurance policy for farm workers.
10. Village level institutions and dairy farmer associations should be ensure that organic farming programme reaches the maximum number of farmers in rural and urban areas.

Issues for Research

There is potential for on-station and on-farm research in all the dimensions of organic livestock production, *viz.* breeding, feeding, disease control, management, processing, marketing, socio-economic and ethical aspects, it being a virgin area as far as India is concerned. The socio-economic investigations concerning the acceptance of organic livestock production and economics of organic livestock farming may also be taken up by the social scientists engaged in livestock sector. The availability of organic feed and fodder among others may be one big limiting factor in initiating any research programme on organic livestock production since the experimental animals need to be fed at least 80 per cent of the feed and fodder grown organically. The comparative studies on conventional v/s organic v/s traditional animal husbandry along various dimensions needs to be carried out. Alternatives to conventional treatment methods, *viz.* plant based products, homeopathy, and other traditional practices need proper using under organic production systems (Mukherjee and Chander 2005). Use of agrochemical and veterinary drugs in different agro-climatic zones, may be determined so as to identify the potential areas and species of livestock for organic livestock farming promotion. The location specific organic livestock production demonstration farms need to be established to generate awareness on organic animal husbandry practices. The ICAR institutes, SAUs and KVKs may be best suited for this purpose. A coordinated interdisciplinary project on network mode is urgently needed to initiate research work in the area of organic livestock production. The ICAR has already launched this kind of project but it is focused on crops. Therefore, a similar project is strongly recommended in the area of organic animal husbandry.

CONCLUSION

Animals are central to organic farming, and in fact, organic farming is not even possible or sustainable without livestock maintained on the organic holdings. But very little work has so far been done in the area of organic animal husbandry in India. This fact makes it imperative that organic animal husbandry is paid due attention by the policy making bodies, research institutions, the SAUs and other development agencies involved with the R and D work on organic farming. Therefore, appropriate steps are required to be taken urgently towards the development of organic animal husbandry in India.

REFERENCES

Banerjee, A. (2001). The organic dairy and food industry – a paradigm shift. Indian Dairyman, 31-33.

Bhattacharya, P. (2004). Organic food Production in India-Status, Strategy and Scope. Agribios (India), Jodhpur, 01-182.

Bhattacharya, P. and Chakraborty, G. (2005). Current Status of Organic farming in India and Other Countries. Indian Journal of Fertilizers, 111-123.

Chander, M. (2001). Organic milk in India: Looking beyond tomorrow. Indian Dairyman, 35-39.

Chander, M. and Mukherjee, R. (2005). Organic animal husbandry: concept, status and possibilities in India-A review. Indian Journal of Animal Sciences, 1460-1469.

Pflimlin, A. and Kempl, M. (2002). Trends of the organic dairy industry in some European countries. James Rencountres author des Recherches Sur Les Ruminants. Paris, France, 2015-2018.

Ram, C. and Bhavadason, M. (2001). Organic milk. Indian J. of Dairy Bio Sciences, 105-107.

Shandilya, S. K., Singh, A. P., Mishra, A., Ojha, B. and Roy, P. (2017). Chemical Composition of Organically Vis-a-Vis Conventionally Grown Fodders. International Journal of Animal and Veterinary Sciences, 04, 30-33.

Shandilya, S., Chatterjee, A. and Roy, P. (2015). Productive and reproductive performance of cows reared under organic vis-à-vis conventional management system in West Bengal. Indian Journal of Animal Sciences, 1263-1264.

Shandilya, S., Singh, A., Ojha, B., Mishra, A., Jaiswal, M. and Jaiswal, S. K. (2019). Effect of organic production system on productive and reproductive performance of cattle. Indian Journal of Animal Research, 1-4.

https: //www.apeda.gov.in/;https: //www.usda.gov/our-agency/about-usda;

https: //ifoam.bio/why-organic/shaping-agriculture/four-principles-organic;

https: //www.fao.org/organicag/oa-faq/oa-faq1/en/

Chapter 23

Livestock and its Role in Organic Farming

Anuj Kumar Gautam[1]*, Rajanish Chandra Mishra[2] and Rajkumari[3]

[1]Subject Matter Specialist (Animal Science),[2]Subject Matter Specialist (Plant Protection), [3]Subject Matter Specialist (Home Science),
Krishi Vigyan Kendra, Jalaun, Banda University of Agriculture and Technology, Banda, Uttar Pradesh
**e-mail: ak.gautam630@gmail.com*

ABSTRACT

Food quality and safety are the two important factors that have gained ever-increasing attention in general consumers. Conventionally grown foods have immense adverse health effects due to the presence of higher pesticide residue, more nitrate, heavy metals,hormones, antibiotic residue. Moreover, conventionally grown foods are less nutritious and contain lesser amounts of protective antioxidants. In the quest for safer food, the demand for organically grown foods has increased during the last decades due to their probable health benefits and food safety concerns. Organic food production is defined as cultivation without the application of chemical fertilizers and synthetic pesticides or genetically modified organisms,growth hormones, and antibiotics. The popularity of organically grown foods is increasing day by day owing to their nutritional and health benefits. Organic farming also protects the environment and has a greater socio-economic impact on a nation. India is a country that is bestowed with indigenous skills and potentiality for growth in organic agriculture. Although India was far behind in the adoption of organic farming due to several reasons, presently it has achieved rapid growth inorganic agriculture and now becomes one of the largest organic producers in the world. Therefore,organic farming has a great impact on the health of a nation like India by ensuring sustainable development.

***Keywords**: Nutrient cycling, Pasture, Food quality, Livestock, Poultry manure.*

INTRODUCTION

Food quality and safety are two vital factors that have attained constant attention in common people. Growing environmental awareness and several food hazards have substantially decreased the consumer's trust towards food quality in the last decades. Intensive conventional farming can add contamination to the food chain. For these reasons, consumers are quested for safer and better foods that are produced through more ecologically and authentically by local systems. Organically grown food and food products are believed to meet these demands (Rembialkowska, 2007).In recent years organically grown foods have become one of the best choices for both consumers and farmers. Organically grown foods are part of go green lifestyle. But the question is that what is meant by organic farming? (Chopra *et al.*, 2013).

Organic farming as an ecological production management system that promotes and enhances biodiversity, biological cycles and soil biological activity (Nourthbourne, 2003). It is based on minimal use of off-farm inputs and on management practices that restore, maintain and enhance ecological harmony (Winter and Davis, 2006). Organic agriculture is an integrated production management system which promotes and enhances agro-ecosystem health, including biodiversity, biological cycles and soil biological activity (FAO/WHO Codex Alimentarius Commission, 2007). It emphasizes the use of natural inputs and the removal of synthetic fertilizers and pesticides.

According with these Guidelines of Organically Food Produce of the Codex Alimentarius, an organic production system is designed to:

1. Enhance biological diversity within the whole system
2. Increase soil biological activity
3. Maintain long-term soil fertility
4. Recycle wastes of plant and animal origin in order to return nutrients to the soil, thus minimizing the use of non-renewable resources
5. Rely on renewable resources in locally organized agricultural systems
6. Promote the healthy use of soil, water and air as well as minimize all forms of pollution that may result from agricultural practices;
7. Promote the careful processing methods agricultural products in order to maintain the organic integrity and vital qualities of the product at all stages;
8. Become established on any existing farm through a period of conversion, the appropriate length of which is determined by site-specific factors such as the history of the land, type of crops and livestock to be produced.

Why Organic Agriculture?

The goal of organic agriculture is to contribute to the enhancement of sustainability. But what does sustainability mean? In the context of agriculture, sustainability refers to the successful management of agricultural resources to satisfy human needs while at the same time maintaining or enhancing the quality of the environment and conserving natural resources for future generations. Sustainability in organic farming must therefore be seen in a holistic sense, which includes ecological, economic and social aspects.

Organic crop husbandry is based on practices outlined in the various organic standards that have been developed and documented in many countries. The most widely recognised organic standards are those published by the International Federation of Organic Agriculture Movements (IFOAM 2002). Organic farming methods emphasise the use of internal on-farm inputs rather than externally sourced inputs to achieve essential soil fertility, nutrient management and plant protection goals. Self-regulation within an agroecosystem, multi-year management cycles and a focus on prevention rather than reaction, are key principles in organic farming that underpin organic plant production.

India is home to 30 per cent of the total organic producers in the world, but accounts for just 2.59 per cent (1.5 million hectares) of the total organic cultivation area of 57.8 million hectares, according to the World of Organic Agriculture 2018 report. Pandey and Sengupta (2018).

Integrating animals into a farm help creating a closed or semi-closed system where energy and nutrients are recycled (Figure 23.1). Animals can convert non-edible biomass (*e.g.* grass, straw, kitchen waste) into food, while increasing soil fertility with their manure.

Organic animal husbandry means not only feeding organic food and avoiding synthetic food additives and synthetic medicines (*e.g.* antibiotics, grow hormones) but also focusing on satisfying the various needs of the farm animals. Good health and welfare of the animals are among the main objectives. Many farm animals have a multi-functional role (Figure 23.2):

1. Produce dung which is of great importance for soil fertility.
2. Yield products such as milk or eggs for sale or own consumption continuously.
3. Recycle by-products such as straw or kitchen waste.
4. Serve as draught animals for tillage or transport.
5. Produce meat, hides, feathers, horns *etc.*
6. Serve as an investment or a bank.

Figure 23.1: Animals Integrated into the Farm Showing the Flow of Fodder, Dung and Products.

7. Help in pest control (*e.g.* dugs) and weed management (*e.g.* grazing on barren fields).
8. Produce young stock for breeding or sale.

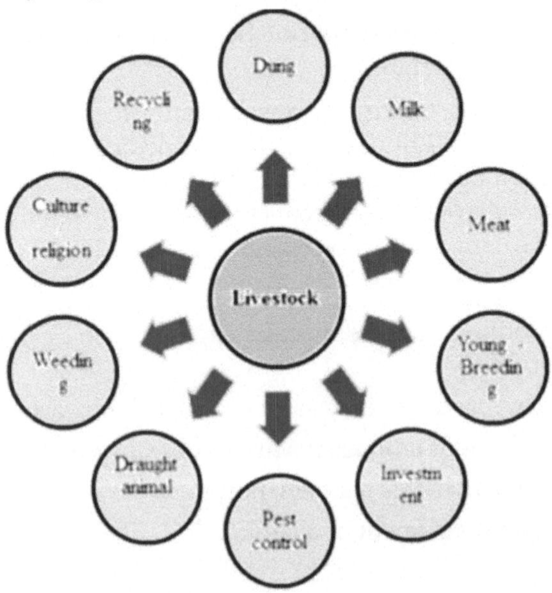

Figure 23.2: Importance of Livestock in Organic Farming.

Organic farmers try to achieve healthy farm animals which can produce satisfyingly over a long period of time. To achieve this goal, various needs of farm animals have to be considered:

1. Fodder in adequate quality and quantity; for non-ruminants: diversity in fodder is usually required.
2. Sufficient access to clean drinking water.
3. Clean sheds of sufficient size and with adequate light and fresh air.
4. Sufficient freedom to move around and perform their natural behaviour.
5. Healthy conditions and veterinary follow-up, when needed.
6. Sufficient contact with other animals, but no stress due to overcrowding.
7. For herd animals: an appropriate age and sex distribution within the herd.

Scenario of Livestock in India

Total livestock population in India is 536.76 million out of which 193.46 million are total cattle population, 109.85 million are total buffalo population, 148.88 millionare total goat population and 74.26 millionare total sheep population *etc.* (Livestock census 2019).Cow dung in India is the major source of organic fertilizer production and used as a co-product in agriculture, such as manure, biofertiliser, biopesticides, pestrepellent and as a source of energy (Dhama *et al.*, 2005). On the other hand farmer also gains 13.87 metric tons of organic fertilizer per year from the biogas plant. This co-production of biofertiliser also allow farmer to recover the initial investment for setting up of a biogas plant (Sharma 2011). As per ayurveda, it can also act as a purifier for all the wastes in the nature (Randhwa and Khullar 2011).In our country, 69.9 per cent population resides in rural areas (The Hindu 2011), where cow (*Bos indicus*) is major cattle and generates 9–15 kg dung/day (Werner *et al.*, 1989; Brown 2003). People in Indian villages use cow dung for cooking purpose by direct burning. It is also used in plastering of walls and floor in rural houses for providing insulation during winter and summer. Accordingly, different usage of cow dung by village peoples reflect the native knowledge associated with it. It also depict that cow plays an important role in village economy and has high socio-economic value (Dhama *et al.*, 2005).

Importance of Livestock

1. Role of Livestock in Energy

According to the United Nations Food and Agriculture Organisation (FAO), the animal waste on this planet produces around 55–65 per cent methane, which upon release in the atmosphere can affect global warming 21 times higher than the rate CO_2 does. Biogas, a mixture of different gases produced by anaerobic fermentation of organic matter from methanogenic bacteria, mainly constitutes methane (50–65 per cent) and CO_2 (25–45 per cent) (Sharma 2011). One kilogram of cow manure can produce 35–40 liter of biogas when mixed with equal amount of water with hydraulic retention time (HRT) of 55–60 days maintained at an ambient temperature of 24–26 °C (Kalia and Singh 2004). Li *et al.* (2009) reported 67 ml/g methane yield from anaerobic digestion of cow manure, whose total and volatile solids were 23.4 and 13.8 g/l, respectively.

Green bacteria such as *Pseudomonas* sp., *Azotobacter* sp. and other purple sulphur or purple non-sulphur bacteria are known to produce maximum amount of methane gas in comparison to other photosynthetic bacteria present in cow dung (Rana *et al.*, 2014). The optimum production of biogas depends upon mesophilic (32–38 °C) and thermophilic (50–55 °C) temperature range (Kashyap *et al.*, 2003). Cow dung is the major source of biogas or gobar gas production in India. Cow dung generated from 3–5 cattle/day can run a simple 8–10 m^3 biogas plant which is able to produce 1.5–2 m^3 biogas per day which is sufficient for the family 6–8 persons, can cook meal for 2 or 3 times or may light two lamps for 3 h or run a refrigerator for all day and can also operate a 3-KW motor generator for 1 h (Werner *et al.*, 1989).

2. Role of Livestock as Enhancer in Agriculture

Human population is increasing worldwide giving rise to intensive farming system and unsuitable cropland management that ultimately results in reduced soil fertility (Onwudike 2010; Bedada *et al.*, 2014). Many disadvantages of widespread use of chemical fertilizers include increase in soil acidity, mineral imbalance and soil degradation (Kang and Juo 1980; Ayoola and Makinde 2008). Due to awareness farmers now a day do not prefer chemical fertilisers (Bedada *et al.*, 2014).In composting,microorganisms decompose organic substrate aerobically into carbon dioxide, water, minerals and stabilized organic matter (Bernal *et al.*, 2009; Kala *et al.*, 2009; Vakili *et al.*, 2015). Compost is added into the soil to improve nutrients and water-holding capacity (Arslan *et al.*, 2008;Vakili *et al.*, 2015). Cow dung as the undigested residue of consumed food material being excreted by herbivorous bovine animal species. When mixture of cow dung and urine in the ratio of 3:1, it mainly consists of lignin, cellulose and hemicelluloses. It also contains 24 different minerals like nitrogen, potassium, along with trace amount of sulphur, iron, magnesium, copper, cobalt and manganese. The indigenous Indian cow also contain higher amount of calcium, phosphorus, zinc and copper than the cross-breed cow (Garg and Mudgal 2007; Randhawa and Kullar 2011). Cow dung harbours a rich microbial diversity, containing different species of bacteria (*Bacillus* spp., *Corynebacterium* spp. and *Lactobacillus* spp.), protozoa and yeast (*Saccharomyces* and *Candida*) (Nene 1999; Randhawa and Kullar 2011). Sawant *et al.* (2007) have isolated many different bacterial genera such as *Citrobacter koseri, Enterobacter aerogenes, Escherichia coli, Klebsiella oxytoca, Klebsiella pneumoniae, Kluyvera* spp., *Morgarella morganii,*

Pasteurella spp., *Providencia alcaligenes, Providencia stuartii* and *Pseudomonas* spp. from cow dung. Recently, researchers observed that addition of cow dung to biomass generated from palm oil in the ratio of 3:1 significantly improved the compost quality with respect to various parameters such as pH, electrical conductivity and C: N ratio (Vakili *et al.*, 2015).Thus, cow dung may not only act as a substitute for chemical fertilisers because it supplements organic matter, but also as a conditioner for soil (Garg and Kaushik 2005; Yadav *et al.*, 2013; Belanger *et al.*, 2014).Thermotolerant *Bacillus subtilis* strains have been recovered from cow dung with great potential in phosphate solubilisation. These Bacillus strains also possessed antagonistic activities against plant pathogens along with production of growth regulators. The findings are significant as isolated bacterial strains being thermotolerant may possibly be used as bio-inoculant in agriculture of tropics where temperature during summer rises up to 42–45°C (Swain *et al.*, 2012).

3. In Nutrient Cycling

Nitrogen fixed by leguminous plants and different nutrients devoured by farm animals amid brushing are come back to soil through dung and urine. Overseen painstakingly, farm animals and manures can assume an imperative part in nutrient cycling on the organic farming.

In feedlots, it is important to store and discard manure and urine in a naturally acceptable way. Excreta contain several nutrients (including nitrogen, phosphorus and potassium) and organic matter, which are important for maintaining soil structure and fertility. Stubble in the fields and crop residues are important sources of forage in smallholder systems. Lower mature leaves stripped from standing crops, plants thinned from cereal stands and vegetation on fallow fields offer additional fodder resources related to food cropping. When animals consume vegetation and produce dung, nutrients are recycled more quickly than when the vegetation decays naturally. Grazing livestock transfer nutrients from range to cropland and concentrate them on selected areas of the farm.

4. In Weed Control

Farm animals are utilized widely for weed control on natural farming. For example, they can graze down weeds either before sowing a crop or after crop establishment for weed control and to enhance tillering. Crops can be selected for their palatability. Farm animals specifically touch out weeds and keep away from the less palatable fodder.

5. In Pasture and Crop Establishment

Farm animals can help with planning the ground for planting. For instance, they can help with stubble management by brushing and trampling the stubble.

6. In Insect and Disease Control

Fodder part of the mixed cropping system builds a critical fertility and structure building phase intorotations and intrudes the potential for the development of insects and disease.

7. In Land Preparation for Cropping

Farm animals such as pigs can 'plough' rough or new land before planting vegetables or grains, reducing tillage and weed control costs.

8. In Soil Water Storage

It results in greater soil water storage capacity, mainly because of biological aeration and the increase in the level of organic matter.

Different Forms of Manures

1. Farmyard Manure

Farmyard manures alludes to the disintegrated blend of dung and urine of livestock alongside litter and left finished material from roughages or pasture nourished to the cows. On an average well decomposed farmyard manure contains 0.5 percent N, 0.2 per cent P_2O_5 and 0.5 per cent K_2O.

2. Sheep and Goat Manure

Excreta of sheep and goats contain higher nutrients than cattle manure and compost. On an average,the manure contains 3 per cent N, 1 per cent P_2O_5 and 2 per centK_2O. It is applied to the field in two ways. Generally sheep or goat shed sweepings are placed in pits for decomposition and later used for field application. The nutrients present in the urine are squandered in this strategy. The second method is sheep penning, where in sheep and goats are kept overnight in the field and urine and dung added to the soil is incorporated to a shallow depth with harrow or cultivator.

3. Poultry Manure

The droppings of birds ferment rapidly. If left exposed, 50 percent of its nitrogen is lost within 30 days. It contains more nitrogen and phosphorus compared to other bulky organic manures. The average nutrient content is 3.03 per cent N; 2.63 percent P_2O_5 and 1.4 per cent K_2O.Other concentrated organic manures: Blood dinner when dried and powdered can be utilized as manure. The meat of dead animals is dried and changed over into meat meal which is a decent source of nitrogen. Nutrient content of concentrated organic manures is given as follows.

Use of Cow Dung in Agriculture in Different Forms

1. As Vermicompost

It is the product of the decomposition process using various species of worms, usually red wigglers, white worms and other earthworms to create a mixture of decomposing vegetable or food waste, bedding materials and vermicast. Vermicast is the end-product of the breakdown of organic matter by earthworms. These castings have been shown to contain reduced levels of contaminants and a higher saturation of nutrients than the organic materials before vermicomposting. Vermicompost contains water-soluble nutrients and is an excellent, nutrient-rich organic fertilizer and soil conditioner.

Containing water-soluble nutrients, vermicompost is a nutrient-rich organic fertilizer and soil conditioner in a form that is relatively easy for plants to absorb. Worm castings are sometimes used as an organic fertilizer. Because the earthworms grind and uniformly mix minerals in simple forms, plants need only minimal effort to obtain them. The worms digestive systems create environments that allow certain species of microbes to thrive to help create a living soil environment for plants.

2. As *Jeevamrit*

Jeevamrit is the traditional Indian bio pesticide and organic manure that is prepared by the unique technique of fermentation of the combined mixture of cow dung, cow urine, jaggery, pulses flour, soil and water. Not only it is cost effective but it is beneficial for both plants and soil. Farmers who spend lots of money on fertilizers and pesticides can save their money and use this amazing traditional medicine for plants. It is 100 per cent organic and has no harmful effects on soil health. It is a rich source of Nitrogen, Potassium and Phosphorus. It also contains all other micronutrients responsible for plant growth and development. It is completely organic and provides all the nutrients required for plant growth and it also prevents plants from pest and diseases. Unlike other organic manure that takes months to be prepared, you can prepare jeevamrit within a week. It helps to maintain the pH of the soil, improves aeration, increase beneficial bacteria, applicable to all plants and much more.

Future Prospects of Organic Farming in India

India is an agriculture-based country with 67 per cent of its population and 55 per cent of manpower depending on farming and related activities. Agriculture fulfills the basic needs of India's fastest-growing population accounted for 30 per cent of total income. Organic farming has been found to be an indigenous practice of India that practiced in countless rural and farming communities over the millennium. The arrival of modern techniques and increased burden of population led to a propensity towards conventional farming that involves the use of synthetic fertilizer, chemical pesticides, application of genetic modification techniques, *etc.*

Even in developing countries like India, the demand for organically grown produce is more as people are more aware now about the safety and quality of food, and the organic process has a massive influence on soil health, which devoid of chemical pesticides. Organic cultivation has an immense prospect of income generation too (Bhardwaj and Dhiman, 2019). The soil in India is bestowed with various types of naturally available organic nutrient resources that aid in organic farming (Adolph and Butterworth, 2002; Reddy, 2010; Deshmukh and Babar, 2015).

CONCLUSION

As we know food quality and safety are two vital factors that have attained constant attention in common people. For these reasons, consumers are quested for safer and better foods that are produced through organically grown food and food products. To implications livestock play an important role. Integrated crop-livestock farming system offers scope to promote organic agriculture; and carry-over of carbon and nutrients from one cropping season to the next. Farm holders according to their resource availability if mixed farming is followed they can improve their livelihoods and standard of living. We know cow dung host a wide variety of microorganisms varying in individual properties. Exploitation of cow dung microflora can contribute significantly in sustainable agriculture and energy requirements. It is one of the bioresources of this world which is available on large scale and still not fully utilised. Another exciting area of research for future studies is developing microbial enzymes and antimicrobials. The production of enzymes by microorganisms from this cheap bioresource can find wide applications in various fields such as agriculture, chemistry and biotechnology. The application of cow dung microflora with considerable antimicrobial potential can result in the promotion of human health; however, comprehensive screening of these microorganisms for the production of antibacterial, antifungal and antiviral metabolites needed to be investigated. In this way, cow dung may be considered as an easily available bioresource that holds a great potential for sustainable development in the near future.

REFERENCES

Adolph, B., Butterworth, J. (2002). Soil fertility management in semi-arid India: its role in agricultural systems and the livelihoods of poor people. Natural Resources Institute, UK.

Arslan EI, Öbek E, Kirba S, Pek U, Topal M (2008) Determination of the effect of compost on soil microorganisms. Int J SciTechnol 3: 151–159

Ayoola OT, Makinde EA (2008) Performance of green maize and soil nutrient changes with fortified cow dung. Afr J Plant Sci 2: 19–22

Bedada W, Karltun E, Lemenih M, Tolera M (2014) Long-term addition of compost and NP fertilizer increases crop yield and improves soil quality in experiments on smallholder farms. AgricEcosyst Environ 195: 193–201.

Belanger G, Rochette P, Chantigny M, Ziadi N, Angers D, Charbonneau E, Pellerin D, Liang C (2014) Nitrogen availability from dairy cow dung and urine applied to forage grasses in eastern Canada. Can J Plant Sci 95: 55–65

Bernal MP, Alburquerque JA, Moral R (2009) Composting of animal manures and chemical criteria for compost maturity assessment—review. BioresourTechnol 100: 5444–5453.

Bhardwaj, M., Dhiman, M. (2019). Growth and performance of organic farming in India: what could be the future prospects? Journal of Current Science, 20: 1–8.

Brown RC (ed) (2003) Biorenewable resources—engineering new products from agriculture. Lowa state press, London.

Chopra, A., Rao, N.C., Gupta, N., Vashisth, S. (2013).Come sunshine or rain; organic foods always on tract: a futuristic perspective. International Journal of Nutrition, Pharmacology Neurological Diseases, 3: 202–205.

Deshmukh, M.S., Babar, N (2015). Present status and prospects of organic farming in India. European Academic Research, 3: 4271–4287.

Dhama K, Chauhan RS, Singhal L (2005) Anti-cancer activity of cow urine: current status and future directions. Int J Cow Sci 1: 1–25.

FAO and WHO(2007) Codex Alimentarius: organically produced food. 3rd edition. Retrieved from: http: //www.codexalimentarius.org/standards/ thematic-compilations/en/

Garg AK, Mudgal V (2007) Organic and mineral composition of Gomeya (cow dung) from Desi and crossbred cows—a comparative study. Int J Cow Sci 3: 1–2

Garg VK, Kaushik P (2005) Vermistabilization of textile mill sludge spiked with poultry droppings by an epigeic earthworm Eiseniafoetida. BioresourTechnol 96: 1063–1071

IFOAM 2002.Final Draft (2002) Basic Standards for Organic Production and Processing.International Federation of Organic Agriculture Movements, Tholey-Theley, Germany.

Kala DR, Rosenani AB, Fauziah CI, Thohirah LA (2009) Composting oil palm wastes and sewage sludge for use in potting media of ornamental plants. Malays J Soil Sci 13: 77–91

Kalia A, Singh S (2004) Development of a biogas plant. Energy Sources 26: 707–714

Kang BT, Juo ASR (1980) Management of low-activity clay soils in Tropical Africa for food crop production. In: Terry ER, Oduro KA, Caveness F (eds) Tropical root crops: research strategies for the 1980s. IDRC, Ottawa, p 129–133.

Kashyap DR, Dadhich KS, Sharma SK (2003) Biomethanation under psychrophilic conditions: a review. BioresourTechnol 87: 147–153

Li R, Chen S, Li X (2009) Anaerobic co-digestion of kitchen waste and cattle manure for methane production. Energy Source Part A Recovery Util Environ Eff 31: 1848–1856

Livestock Census (2019) All India Report.Ministry of Agriculture Department of Animal Husbandry Dairying and Fisheries, Krishi Bhawan, New Delhi.

Nourthbourne, C.J., 5th Lord. (2003). Look to the Land, 2nd Rev Spec edn.Sophia Perennis, Hillsdale, NY; First Ed. 1940.J.M. Dent and Sons.

Onwudike SU (2010) Effectiveness of cow dung and mineral fertilizer on soil properties, nutrient uptake and yield of sweet potato (Ipomoea batatas) in Southeastern Nigeria. Asian J Agric Res 4: 148–154

Rana G, Mandal T, Mandal NK (2014) Generation of high calorific fuel gas by photosynthetic bacteria isolated from cow dung. Int J Res 1: 115–128

Randhawa GK, Kullar JS (2011) Bioremediation of pharmaceuticals, pesticides, and petrochemicals with gomeya/cow dung. ISRN Pharmacol. doi: 10.5402/2011/362459

Reddy S.B. (2010). Organic farming: status, issues and prospects—a review. Agricultural Economics Research Review, 23: 343–358.

Rembialkowska, E. (2007). Quality of plant products from organic agriculture. Journal Science of Food and Agriculture, 87: 2757–2762.

Sawant AA, Hegde NV, Straley BA, Donaldson SC, Love BC, Knabel SJ, Jayarao BM (2007) Antimicrobial-resistant enteric bacteria from dairy cattle. Appl Environ Microbiol 73: 156–163

Sharma CK (2011). Biogas—a boon for India. Biofuels 2–3

Swain MR, Laxminarayana K, Ray RC (2012) Phosphorus solubilization by thermotolerant Bacillus subtilis isolated from cow dung microflora. Agric Res 1: 273–279

The Hindu (2011) About 70 per cent Indians live in rural areas: Census report. http: //www.thehindu.com/news/national/about70percentindiansliveinruralareascensusreport/article2230211.ece. Accessed 24 Jul 2015

Vakili M, Zwain HM, Rafatullah M, Gholami Z, Mohammadpour R (2015) Potentiality of palm oil biomass with cow dung for compost production. KSCE J Civil Eng 19: 1994–1999

Werner U, Stöhr U, Hees N (1989) Biogas plants in animal husbandry. Deutsc hesZentrumfürEntwicklungstechnologien-GATE.

Yadav A, Gupta R, Garg VK (2013) Organic manure production from cow dung and biogas plant slurry by vermicomposting under field conditions. Int J Recycl Org Waste Agric 2: 21

Yadava, A. K. (2019). Current Status of Organic Farming and Role of Government Institutions in India. Retrieved from https: / /papers.ssrn.com/sol3/papers. cfm? abstract_id=

Index

C

D

E

F

G

9 789359 191645

Printed at Repro India Ltd.